MATH POWER
for
Secondary Schools

I0489104

Book 1

Luutu Suleiman

-------------------------------------Create Space Imprint -------------------------------------

TABLE OF CONTENTS

Preface

MATHEMATICS POWER FOR SECONDARY SCHOOLS (MPSS) series are specially written for secondary school learners, developing the four years secondary school mathematics course. Book 1 is the first book in the series and it is specially for learners in their first year of secondary school.

Other readership involving secondary school teachers, tutors and learners in Teacher Training Colleges (for both primary and secondary schools), Technical schools and institutes, etc. may find the series selectively important to their learning or class work preparations.

Finally, adult learners who wish to further their knowledge in secondary school mathematics may too find the series important to them.

The purpose of the series is to provide a well structured approach in numerical and analytical secondary school mathematics knowledge and skills.

The organization and presentation is such that all the work is logical, coherent, comprehensive and objectively handled to equally cater for the average and the more able students.

In a systematic manner using clear instructional language for explanatory notes, worked out examples, illustrations, tables, figures, etc. the book allows the reader sufficient involvement in mathematical reasoning, convenient work rate, comprehension, and to stimulate interest in the subject being studied. The concepts are related to real life experiences in order to enhance their methodical treatment as much as possible.

For the learners' practice and assessment, brain storming trial questions, graded exercises, revision exercises and examination type papers are provided.

Suleiman Luutu
Namielus Educational

1 SET CONCEPTS

INTRODUCTION

Sets help us identify things such as objects, numerals, letters or symbols and so on, in groups depending on their respective characteristics or nature. Through understanding how various groups differ or how they are similar, we can easily relate them and hence use the idea for various applications.

1.1 Sets

When you look around your environment, you can see a lot of different and similar things. Such things may include plants, buildings, roads, and many more. The concept of sets is therefore concerned with identifying, sorting and grouping of things following a certain common identity.

1.1.1 Identifying, Sorting and Grouping of things

To know whether some things are similar or not, we must be able to understand what they are and what makes them similar or different from others. By doing this, we are *identifying* them. When we know what each thing is, we can then select them according to what they are; and this means *sorting* them. Having known the similar things, we can then put them together according to their similarity. This is known as *grouping*.

Activity 1.1

1 Collect a number of things around you and name them.

2 Find the different ways you can group the collected items.

3 Clearly list the names of things in each group.

A charismas tree, sunflower, rose and a palm tree are all *plants*; but they can be regrouped according to type. The possible types of plants can be *trees* and *flowers*. See fig. 1.1 on next page.

- The group of *trees* constitutes a *christmas tree* and a *palm tree*.
- And the group of *flowers* constitutes a *sunflower* and a *rose*.

Group of plants

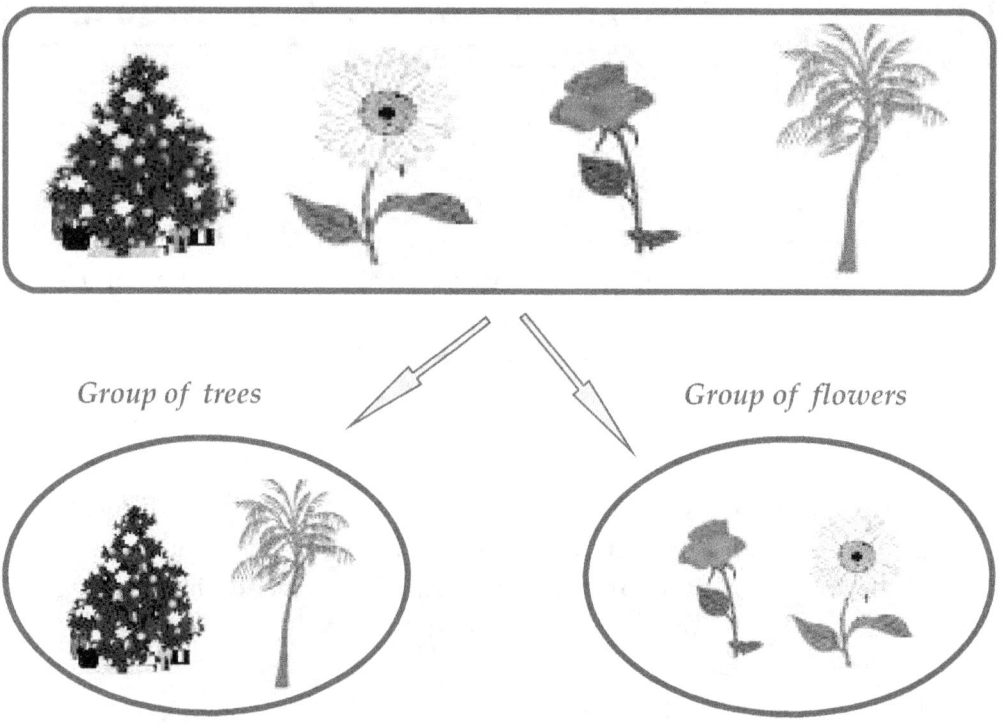

Group of trees

Group of flowers

Fig. 1.1

1.2 Meaning of a set

To group things, we depend on their similarities or differences. A group of similar things will always form a set. We can therefore define a set as a collection of well defined things with something in common. A set can be made of things such as objects, numerals, letters, words, symbols, plants, animals and so on. Special names are used to refer to a number of sets, say, a swarm for bees, a flock for birds, a herd for cattle, etc.

Sets can also be named as follows;

- a set of consonants of the English alphabet,
- a set of months of the year,
- a set of days of the week, etc.

A set is usually denoted a capital letter and items in it shown listed or it is defined with in a pair of curl brackets, { }. If a set A = {a, b, c, d, e, f}; then the letters a, b, c, d, e and f are members of the set A. Members of a set are also called its *elements*.

Sets can be represented using Venn diagrams. Venn diagrams are a pictorial representation of sets with in closed loop boundaries. The loop of a particular set can take the shape of a circle, oval, rectangle, square or any other shape.

Set description: **V** = {vowels of the English alphabet}

Set elements: **V** = {*a, e, i, o, u*}

On a Venn diagram: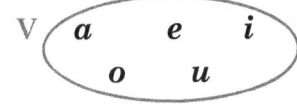

1.3   Description of a set

Members of the same set should obey and follow a similar rule. A set is always labeled as a specific group or simply represented by a letter.

If set A = {0, 2, 4, 6, 8, 10}, then, set A is a set of all even numbers up to and including 10.

Activity 1.2

State the common characteristic (s) of the elements of each of the following sets   **P** = {Chair, Cupboard, Bench, Book shelf, Bed} and

Q = {Rubber, Plastic, Wood, Cork, Paper, Cloth}

1.3.1   Naming of a set and listing of its members

Given a set Q, such that; Q = {A set of Even numbers}, then the members of set Q are listed as; Q = {0, 2, 4, 6, 8, 10, 12, 14, 16 ...}. You should be able to get the name of a set with the help of a common characteristic between or among the members of that set.

Example 1

List all the members of each of the sets given below.

A = {a set of letters that form the word "country"}

B = {a set of the seven colours of the rainbow}

C = {a set of all prime numbers between 4 and 25}

D = {a set of all Vowels of the English alphabet}

Solution

$A = \{c, o, u, n, t, r, y\}$
$B = \{$Red, Orange, Yellow, Green, Blue, Indigo, Violet$\}$
$C = \{5, 7, 11, 13, 17, 19, 23\}$ $D = \{a, e, i, o, u\}$

Example 2

List the members of a set of the first 10 composite numbers.

Solution

A composite number is one with other factors other than 1 and the number itself. If the set of the first 10 composite numbers is C,
then set $C = \{4, 6, 8, 9, 10, 12, 14, 15, 16, 18\}$

Example 3

Name the following sets:
   $P = \{$Lion, Leopard, Chitter, Tiger, Jaguar$\}$
   $Q = \{$Corrolla, Carina, Corsa, Corona$\}$
   $R = \{$Square, Rectangle, Parallelogram, Rhombus, Trapezium, Kite$\}$
   $S = \{$Red, Blue, Green$\}$

Solution

   $P = \{$A set of wild animals of the cat family$\}$
   $Q = \{$A set of Toyota company cars with names starting with letter C$\}$
   $R = \{$A set of four sided plane figures, quadrangles or quadrilaterals$\}$
   $S = \{$A set of primary colours$\}$

Exercise 1a

List down all the possible members of each of the following sets.

1 $P = \{$Polygons with 4 sides$\}$
2 $Q = \{$Solid figures with six faces$\}$
3 $R = \{$East African countries with names starting with consonants$\}$
4 $S = \{$Multiples of 3 which are even$\}$

5   T = {Multiples of 4 between 1 and 100 which are divisible by 6}

Given the members of the following sets state their names;

6 U = {Ream of papers, Clip board, Staples, Pen}

7 V = {Jan, Feb, Mar, Apr, May, Jun, Jul, Aug, Sept, Oct, Nov, Dec}

8 W = {B, C, D, F, G, H, J, K, L, M, N, P, Q, R, S, T, V, W, X, Y, Z}

9 X = {C, L, O, S, E, D}

10 Y = {2, 4, 6, 8, 10, 12 ... }

If set S = {1, 2, 3, 4, 5, 6, 7, 8 ... }: state *true* or *false*.

11 S is a set of numbers grater than zero.

12 S is a set of counting numbers

13 124 is not a member of set S.

14 All prime numbers are members of set S.

15 S is a set of numbers exactly divisible by two.

16 S = { 1, 2, 3 . . . 9, 10, 11 . . . }

1.3.2 Set notations

Symbols are used as a language to represent set terminologies, relations or operations. This system is known as *set notation*.

Member of; ∈

If set V = {*a, e, i, o, u*}, then, members of set V are *a, e, i, o* and *u*. "*a*" is a member of set V; represented as: $a \in V$. "2" is not a member of set V, and this can be represented as: $2 \notin V$.

Number of; n(A)

The notation n(A) means, number of members in set A.

For set P = {2, 3, 5, 7}, then n(P) = 4

Given a set V: 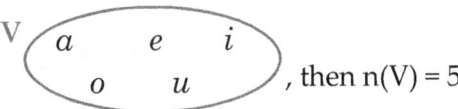 , then n(V) = 5

Contained in; ⊃

If set V = {*a, e, i, o, u*} and another set A = {*a, u*} we realize that all members of set A are also in set V. So, set A is contained in set V and this is represented as A ⊃ V.

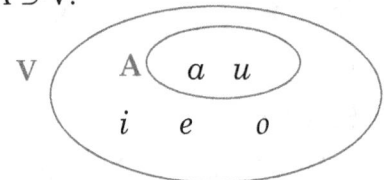

For a set B = {1, 2, 3}, we realize that it is not contained in set V and this is represented as B ⊅ V.

Set V is also known to be the superset for set A. A superset sometimes is also referred to as a universal or a mother set.

Subset of; ⊂

For the set V = {a, e, i, o, u} and A = {a, u}; set A is a subset of set V. This is represented as A ⊂ V. If a set B = {1, 2, 3}, we realized that B is not a sub set of set V, that is, B ⊄ V.

Exercise 1b

Given sets A = {a, b, c, d, e, f} and B = {0, 2, 4, 6, 8}, state *true* or *false* in the following;

1 b ∈ A	5 n(A) = n(B)	9 {2} ⊃ B
2 e ∈ B	6 Ø ⊂ B	10 {a, b, c} ⊃ A
3 c ∈ A	7 A ⊂ B	11 A ⊂ {a, b, c, d, e}
4 2 ∈ A	8 A ⊃ B	12 {2, 4, 8} ⊂ {0, 2, 4, 6, 8}

1.4   Types of sets

1.4.1 Finite and Infinite sets

Finite sets are sets with a countable number of elements. A set of books in my bag, fingers on my palms and birds in the world are examples of finite sets. A set W, of days of the week is finite, because the number of its elements is specific.

W = {Sunday, Monday, Tuesday, Wednesday, Thursday, Friday, Saturday}

Infinite sets are sets with an endless list of elements. A set of all counting numbers, a set of all odd numbers or a set of all composite numbers are examples of infinite sets. A set C, of all counting numbers is infinite because its number of elements is endless. Counting numbers can never be completed, therefore, for a set they can be represented as; $C = \{1, 2, 3, 4, 5, 6 \dots \}$.

While identifying sets, avoid confusing finite and infinite sets in some cases. For example, a set of all living creatures in the world may be confused to be infinite, but it is not. It may be very difficult to count the creatures, but they are countable.

Activity 1.3

1 Do you think a set of primary school children in Uganda is infinite or finite? Why?
2 State whether the cells in a human body are infinite or finite. Explain your argument.

Example 1

Given that;
A = {All even numbers}, **B** = {All prime numbers which are even} and **C** = {Square numbers which are also rectangular numbers}.
State whether each of the sets above is finite or infinite.

Solution

Set A is an infinite set, since it is, **A** = {0, 2, 4, 6 ... }
Set B is a finite set, since it is **B** = {2}
Set C is an infinite set since all square numbers are rectangular, and all numbers are rectangular except the prime numbers.

1.4.3 Empty sets " Ø " or " { } "

An *empty set* is a set which has no members
Example 1

Think of and write down any five empty sets

Solution

Set **A** = {Elephants studying in our classroom}, Set **B** = {Pregnant men}
Set **C** = {Even numbers which are odd}, Set **D** = {Triangles with 4 sides} and
Set **E** = {Multiples of 10 with 2 as the last digit}

A set of rivers in our class is an empty set. This is because we don't have rivers in our class. A set of crying stones is an empty set, since there are no stones which cry. An empty set is also known as a *null set*.

Exercise 1c

1 Think of and write down any five empty sets.

2 State *true* or *false*

  (a) $\varnothing \subset Q$, where Q is any set.

  (b) If A = {All even prime numbers}, then A = { }

  (c) If M = {All prime multiples of 5}, then M = { }

  (d) If T = {All square numbers that are cubic}, then T = { }

3 List any five finite sets

4 List any five infinite sets.

5 State *true* or *false*

  (a) A set of dust particles in the atmosphere is an infinite set.

  (b) A set of birds in the Wild Life Center, Entebbe is a finite set.

  (c) A set of all presidents in Africa is an infinite set.

  (d) A set of human cells in the whole world is a finite set.

  (e) A set of all natural numbers is an infinite set.

1.5 Relationships between sets

Sets may be categorized according to how they relate to each other.

1.5.1 Equal " = " and Equivalent " ≡ " sets

Equal sets are sets with similar members. Therefore, two sets are said to be equal if they contain exactly the same members.

For example, if Set A = {1, 3, 8, 9}, and Set B = {3, 8, 9, 1}, we can say that set A equals to set B, represented as; Set A = Set B.

If set P = {letters in the word queen} and set Q = {q, u, e, n}, we still conclude that Set P = Set Q. Note that letter "e" is not repeated set Q.

Example 1

State whether the sets below are equal or not.

(a) **A** = {1, 3, 5, 7} and **B** = {1, 3, 5, 7}

(b) **C** = {0, 2, 4, 6} and **D** = {0, 2, 4}

(c) **E** = {1, 2, 3 ... } and **F** = {1, 2, 3, 4, 5 ... }

(d) **G** = {a, e, i, o, u} and **H** = {Vowels of the English Alphabet}

Solution

(a) Set A = Set B, since all members in A are also in B.

(b) Set C ≠ Set D, since 6 is only in C and not in D, (≠ means "not equal to").

(c) Set E = Set F. They both start from 1 and are continuous.

(d) Set G = Set H. Set H is the naming of Set G.

Equivalent sets are sets with an equal number of members or elements. The elements may not necessarily be the same. Therefore all equal sets are also equivalent, but not all equivalent sets are equal.

If P = {1, 4, 8, 12} and Q = {3, 8, 4, 9}; then P is equivalent to Q, represented as P ≡ Q. The number of elements in P is equal to the number of elements in Q, that is, four elements in each set. If U = {1, 3} and V = {0, 2, 4}, then U ≢ V (U is not equivalent to V).

Example 2

State whether the following pair of sets is equivalent or not.

(a) Set **A** = { △, □, ◁, ⬠ } and Set **B** = { L, M, N, P }

(b) Set **C** = {3, 6, 9, 12} and Set **D** = {9, 12, 15, 18}

(c) Set **E** = {0, 2, 4, 6, 8} and Set **F** = {0, 2, 4, 6, 8}

(d) Set **G** = {1, 2, 3} and Set **H** = {1, 2, 3, 4}

Solution

(a) Set A ≡ Set B, since they have the same number of elements.

(b) Set C ≡ Set D, since they have the same number of elements.

(c) Set E ≡ Set F, (Also E = F)

(d) Set G ≢ H, since they have a different number of elements.

Activity 1.4

Given that, Set P = {the first 5 prime numbers}, and
        Set Q = {the first 5 odd numbers}. State whether the sets P and Q are equal, equivalent or neither of these and give a reason (s) why?

Exercise 1d

Given that, Set A = {1, 3, 6, 10},   Set B = {0, 2, 4, 6, 8},

Set C = $\left\{ \triangle , \square , \diamond , \pentagon \right\}$,

Set D = {1, 2, 3, 4 ...}   ,   Set E = { 1, 3, 6, 10, 15 ...} ,

Set F = {1, 3, 6, 10, 15} and   Set G = {L, M, N, O, P}

State *true* or *false* and give a reason (s) why?

1 Set C = Set A	5 Set F ≡ Set A	9 Set G = Set C
2 Set D ≡ Set G	6 Set B ≡ Set G	10 Set B ≠ Set F
3 Set D = Set E	7 Set F = Set A	11 Set A ≢ Set C
4 Set B = Set G	8 Set D ≢ Set E	12 Set E ≠ Set F

1.5.2 Universal set, ξ and subsets, ⊂

A universal set is one which exhausts all the members of the same characteristics, hence being with all the elements for a particular description. Given a universal set ; ξ = {0, 1, 2, 3, 4, 5, 6, 7, 8, 9, 10}, of whole numbers, we may obtain the following subsets from it.

Subsets	*Members of the subsets*
E = {Even numbers}	E = {0, 2, 4, 6, 8, 10}
P = {Prime numbers}	P = {2, 3, 5, 7, 9}
T = {Triangular numbers}	T = {1, 3, 6, 10}

Conclusively, set E, P and T are sub sets of the universal set given before. Study fig. 1.2 on page 11 for the Venn diagram representation.

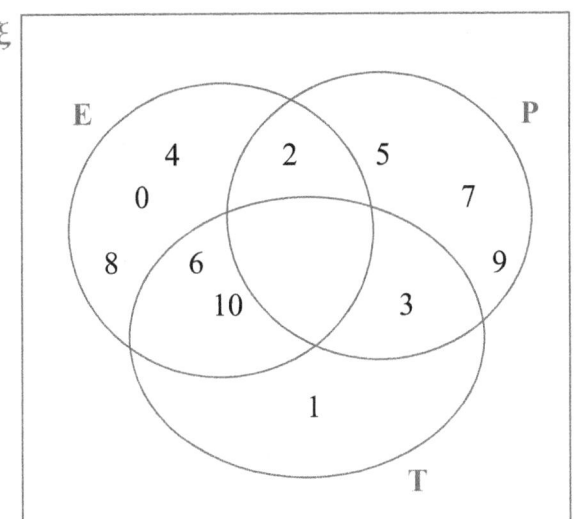

Symbolically;

$E \subset \xi$, $P \subset \xi$ and

$T \subset \xi$

The universal set also contains sets E, P and T.

Fig. 1.2

Always note;

(i) If $X \subset Y$, it means that all members of set X are also in set Y. For example, where Set X = {1, 3} and Set Y = {1, 2, 3, 4}.

(ii) If set R = set S, then $R \subset S$ and $S \subset R$. For example, if set R = {1, 2, 3, 4, 5} and S = R, then set R is a sub set of set S, and set S is a sub set of set R. Therefore; for equal sets, one is a subset of the other.

(iii) Any set is a subset of itself.

(iv) An empty set is a subset of any set.

(v) A proper subset is one which does **NOT** contain **ALL** the members of its mother set and it is not empty.

Example 1

Given the sets; Q = {0, 1, 2, 3, 4, 5, 6, 7, 8, 9}, R = {2, 5, 7}

S = {2, 3, 5, 7, 11}   T = {1, 3, 5, 7}   U = {2, 5, 7}

State *true* or *false:*

1 $R \subset T$ 2 $S \subset Q$ 3 $R \subset U$ 4 $Q \subset U$ 5 $T \subset Q$

Solution

1 False   2 False   3 True   4 False   5 True

Activity 1.5

1 State *true* or *false* for $\emptyset \subset R$, where R is any set, and explain.

2 Give explanations for all the answers in the example 1 above.

Finding subsets for a given universal set

Study the following table carefully.

No. of elements in universal set (n)	Universal set (ξ)	Subsets listed, (s)	No. of Subsets (S)
0	$\xi_0 = \{ \ \}$	$s_1 = \{ \ \}$	1
1	$\xi_1 = \{a\}$	$s_1 = \{ \ \}, \ s_2 = \{a\}$	2
2	$\xi_2 = \{a, b\}$	$s_1 = \{ \ \}, \quad s_2 = \{a\}, \quad s_3 = \{b\},$ $s_4 = \{a, b\}$	4
3	$\xi_3 = \{a, b, c\}$	$s_1 = \{ \ \}, \quad s_2 = \{a\}, \quad s_3 = \{b\},$ $s_4 = \{c\}, \quad s_5 = \{a, b\},$ $s_6 = \{a, c\}, \quad s_7 = \{b, c\},$ $s_8 = \{a, b, c\}$	8
4	$\xi_4 = \{a, b, c, d\}$	$s_1 = \{ \ \}, \ s_2 = \{a\}, \ s_3 = \{b\},$ $s_4 = \{c\}, \ s_5 = \{d\}, \ s_6 = \{a, b\},$ $s_7 = \{a, c\}, \quad s_8 = \{a, d\},$ $s_9 = \{b, c\}, \quad s_{10} = \{b, d\},$ $s_{11} = \{c, d\}, \quad s_{12} = \{a, b, c\},$ $s_{13} = \{a, b, d\}, \ s_{14} = \{a, c, d\},$ $s_{15} = \{b, c, d\}, \ s_{16} = \{a, b, c, d\}$	16

Fig. 1.3

Activity 1.6

List all the subsets of $\xi = \{a, b, c, d, e\}$. How many subsets are there?

Have you realized how difficult it is to list all the sub sets of sets with five elements and more? We therefore use a formula to find subsets of a given set if we know the number of elements it has.

No. of subsets, S $= 2^n$, where n is the number of elements in the set.

Example 2

Find the number of subsets for each of the following sets:

$\xi_1 = \{A, B, C\}$ $\xi_2 = \{a, e, i, o, u\}$ and

$\xi_3 = \left\{ \triangle\ \square,\ \Diamond,\ \diagdown,\ \triangle,\ \Diamond \right\}$

Solution

Let the number of subsets be: S

(i) $S_1 = 2^n$, but n = 3   (ii) $S_2 = 2^n$, but n = 5   (iii) $S_3 = 2^n$, but n = 6

$S_1 = 2^3$ $S_2 = 2^5$ $S_3 = 2^6$

$S_1 = 8$ subsets $S_2 = 32$ subsets $S_3 = 64$ subsets

Exercise 1e

Given the sets;

   $A = \{1, 3, 5, 7, 9, 11 \dots 21\}$ $B = \{2, 3, 5, 7, 11\}$

   $C = \{5, 7, 9, 11, 13, 15\}$ $D = \{1, 2, 3, 4, 5, 6, 7, 8, 9 \dots\}$

State *true* or *false*

1 $B \subset C$ 2 $B \subset A$ 3 $C \subset A$ 4 $A \subset D$ 5 $D \subset C$

6 (a) Given the sets below, find the number of subsets for each of the following sets. $A = \{1, 2, 3, 4, 5, 6\}$, $B = \{a, e, i, o, u\}$, and
$C = \{U, V, W, X, Y, Z, P, Q, R, S, T\}$

 (b) If a set has 128 subsets, how many elements does it have?

1.6   Operations on sets

1.6.1 Union of sets; $\cup$

Given the sets P = $\{1, 4, 8, 10\}$ and Q = $\{2, 4, 8, 9, 12\}$, then the union of sets P and Q; $P \cup Q$ is given by a set of all members found in both sets.

Therefore, $P \cup Q = \{1, 2, 4, 8, 9, 10, 12\}$.

On a Venn diagram;

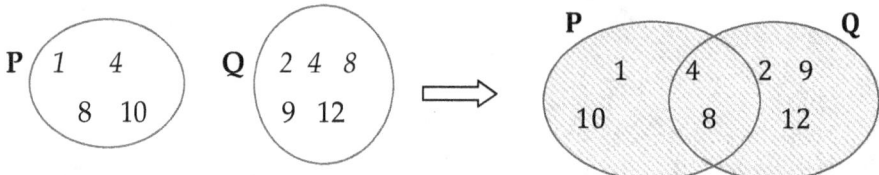

Note that members appearing in both sets are written once (are not repeated) in the union of the sets. For example, 4 and 8 are common for both sets. The shaded area represents the union of sets.

1.6.2 Intersection of sets; ∩

For the sets P and Q, the intersection of such sets is the set containing all the elements which are common in both sets. Given a Set P = {1, 3, 6, 10, 15} and Q = {3, 6, 9, 12, 15}, we see that the elements 3, 6 and 15 are common in both sets, so they represent the intersection of sets P and Q.

Therefore, P∩Q = {3, 6, 15}

On a Venn diagram;

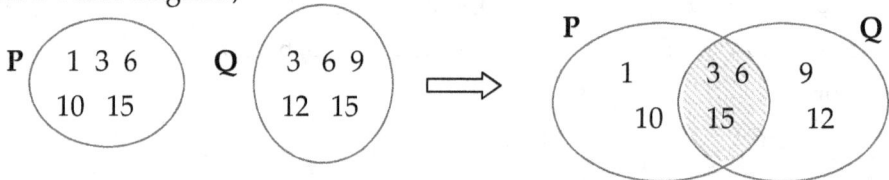

The shaded part represents the intersection of sets P and Q.

1.6.3 Joint and Disjoint sets

If any two sets have atleast a common element, then they are said to be *joint*. If they have no any common elements, then they are said to be *disjoint*. Therefore the intersection of any two disjoint set is an empty set. A ∩ B = { } is a condition for which sets A and B are disjoint.

For sets **A** = {1, 2, 3} and   On the venn diagram;
   B = {4, 6, 8, 10}; being disjoint.

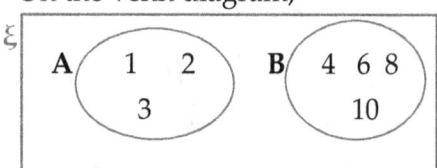

Note that;

The loop for set A does not cross that of set B.

1.6.4 Set difference

A set difference constitutes of elements of a set P, which are not elements of set Q. For set P = {1, 3, 6, 10} and Q = {2, 3, 5, 7}, then P difference Q; P – Q = {1, 6, 10}; (elements in set P and not in set Q).

On the venn diagram;

P – Q (Shaded)   Q – P (Shaded)

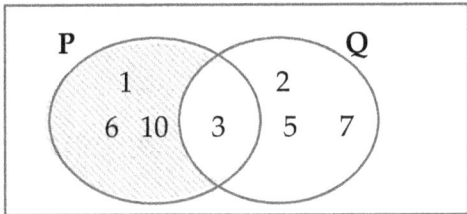   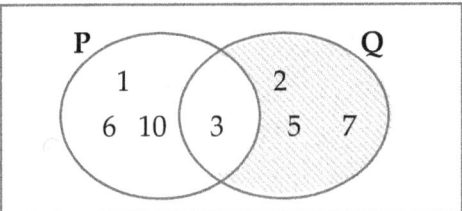

1.6.5 Complement of a set, A'

This refers to the members existing in one set, but not existing in the other. Complement of a set A is denoted A' (A-complement). It means all the members out side Set A, but in the universal set. Given the universal set,

ξ = {5, 10, 15, 20, 25, 30, 35} and set A = {5, 15, 25, 35}, elements which are in the universal set, but not in set A are 10, 20 and 30; therefore,   A' = {10, 20, 30}.

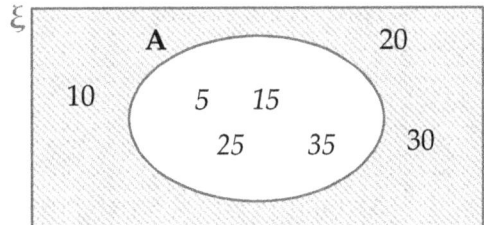

Exercise 1f

1 Given set P = {a, b, d, e, s} and set Q = {2, 4, 6, 8, 9}, state *true* or *false* in the following,

   (i) b ∈ P   (ii) e ∈ Q   (iii) c ∈ P

   (iv) 2 ∈ P   (v) n(P) = n(Q)   (vi) Ø ⊂ Q

   (vii) P ⊂ Q   (viii) P ⊃ Q   (ix) {6} ⊃ Q

   (x) {a, b} ⊃ P

2 Given sets   P = { 1, 2, 3, 4 …},   Q = { 2, 3, 5, 7, 11} and R = { 0, 2, 4, 6, 8}
   (a) Find (i) Q ∪ R   (ii) Q ∩ R   (iii) P ∪ R

(iv) $P \cap R$ (v) $n(Q \cap R)$

(b) State *true* or *false*   (i) $n(Q) = n(R)$ (ii) $R \supset P$

                     (iii) $Q \subset P$ (iv) $R \subset Q$

3 Given sets, $\xi = \{0, 1, 2, 3 \ldots 100\}$, $P = \{2, 3, 5, 7, 11 \ldots 97\}$

    $C = \{4, 6, 8, 9, 10, 12, 13 \ldots 100\}$, $E = \{0, 2, 4, 6, 8, 10, \ldots 100\}$

    $D = \{1, 3, 5, 7 \ldots 99\}$ and $S = \{1, 4, 9, 16, \ldots 100\}$

(a) Find   (i) $n(\xi)$ (ii) $E \cup D$ (iii) $D \cap S$

            (iv) $P \cap E$ (v) $E \cap D$ (vi) $(P \cap C)'$

            (vii) $n(P \cap C)'$ (viii) $(P \cap C)'$ (ix) $n(E \cap S)$

            (x) $n(P)$

(b) State *true* or *false*

    (i) $E \cup D = \xi$ (ii) $n(E) = n(D)$ (iii) $E \cap P = \{2\}$

    (iv) $C \cap P = \{0, 1\}$ (v) $1 \in P$ (vi) $C \supset \xi$

    (vii) $E \cap D = \{\ \}$ (viii) $n(\xi) = 100$ (ix) $E \equiv D$

    (x) $n(P) + n(C) = n(\xi)$

4 Given the sets;

   $A = \{$All square numbers less than 1000$\}$

   $B = \{$All multiples of 3 which are square numbers and less than 1000$\}$

   $C = \{$All cubic numbers less than 1000$\}$

   $D = \{$All square numbers which are cubic and are less than 1000$\}$

(a) List the members of the sets A, B, C and D.

(b) Find (i) $A \cap B$ (ii) $A \cap B \cap C$ (iii) $B \cap D$

          (iv) $(A \cap C)$ (v) $n(B \cap D)$ (vi) $n(A \cap C)$

(c) State *true* or *false*

          (i) $(A \cap C) = D$ (ii) $B \supset A$ (iii) $n(A) = n(C)$

         (iv) $n(A \cap C) = n(D)$ (v) $n(A) > n(B)$ (vi) $B \supset C$

1.7 More Venn diagrams

Venn diagrams are pictorial representations of sets. It involves displaying the elements of the set or the number of these elements.

1.7.1 Representing elements of sets on Venn diagrams

We can indicate specific elements of a set on Venn diagrams. Given that set P = {p, e, n} and set Q = {q, u, i, t, e}, then we can represent these two sets on Venn diagrams as follows:

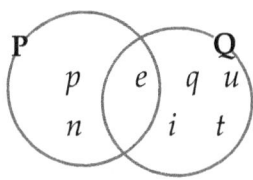

Note that the commas are not shown on the Venn diagrams and that the element *e* is in both sets, so it is written once in a region   shared   by both the sets.

Example 1

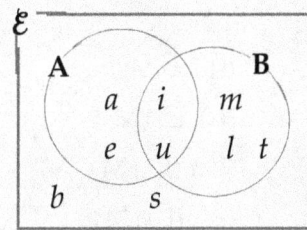

Given the Venn diagram on left;

Find   (a) A   (b) B   (c) ℰ

(d) (A ∪ B)'   (e) A' ∩ B

Solution

(a) A = {a, e, i, u}   (b) B = {i, l, m, t, u}   (c) ℰ = {a, b, e, i, l, m, s, t, u}

(d) (A ∪ B)' = {b, s}   (e) A' ∩ B = {l, m, t}

Example 2

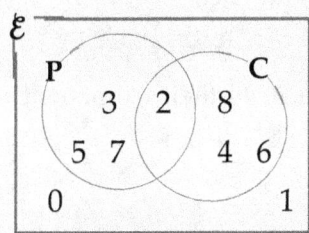

Given the Venn diagram on left;

Find   (a) P   (b) C

(c) P ∩ C   (d) (P ∪ C)'

Solution

(a) P = {2, 3, 5, 7} (b) C = {2, 4, 6, 8}   (c) P ∩ C = {2} (d) (P ∪ C)' = {0, 1}

Exercise 1g

1 Given a universal set $\mathcal{E}$ = {1, 2, 3, 4, 5, 6, 7, 8 . . . 15}, set **L** = {1, 3, 6, 10}, set **M** = {2, 3, 5, 7} and **N** = {3, 9, 12, 15}.

(a) Show the sets on a venn diagram.

(b) Find (i) L ∩ M ∩ N   (ii) (L ∪ M ∪ N)'

2 Given the Venn diagram below;

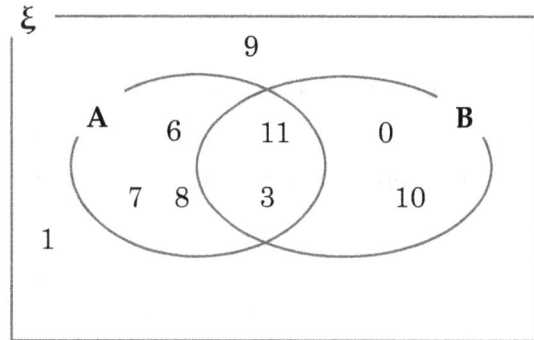

Find the members of

(i) set A

(ii) set B only

(iii) A ∩ B

(iv) (A ∪ B)'

(v) n(ξ)

3 Given the Venn diagram below;

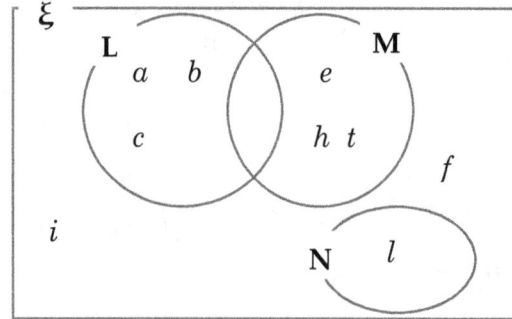

(a) Find,   (i) L ∪ M

(ii) L ∩ M

(iii) (L ∪ M ∪ N)'

(b) Find,   (i) n(L ∩ M)

(ii) n(M ∪ N)'

(iii) n(L)'

1.7.2   Representing the number of elements on Venn diagrams

Given a set L = {a, b, c, d, e, f, g, h, i} and set V = {a, e, i, o, u}, their elements can be shown on Venn diagrams as follows:

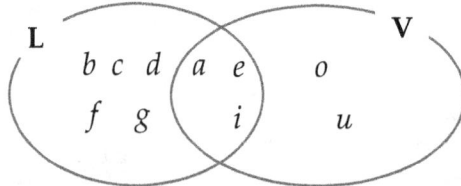

We can however rewrite this as number of elements on Venn diagrams. Follow next page.

Elements of the sets L and V:

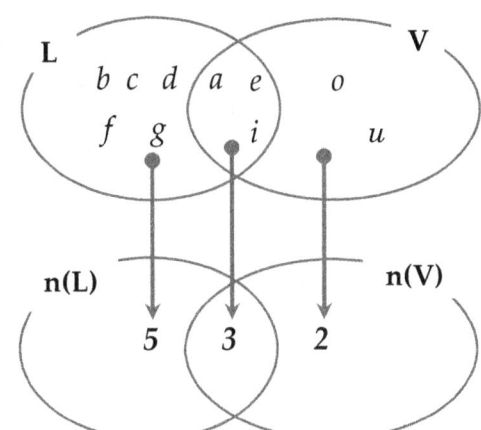

Number of elements for sets L and V:

We can now summarize that;

(i) n(L) $= (5 + 3)$ (ii) n(V) $= (3 + 2)$
          $= 8$ $= 5$

(iii) $n(L \cap V)$ $= 3$ (iv) $n(L \cup V)$ $= (5 + 3 + 2)$
                                              $= 10$

Example 2

Given that set N = {1, 2, 3, 4, 5, 6, 7, 8, 9} and set C = {4, 6, 8, 9, 10}. Represent their number of elements on a Venn diagram and find:

(i) $n(N \cap C)$ (ii) $n(N \cup C)$

Solution

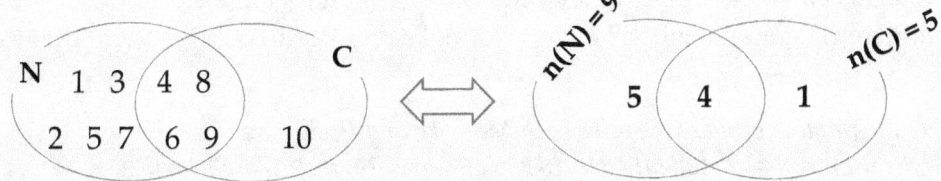

(i) $N \cap C$ = {4, 6, 8, 9}. So, $n(N \cap C) = 4$
(ii) $N \cup C$ = {1, 2, 3, 4, 5, 6, 7, 8, 9, 10}. So, $n(N \cup C) = 10$

Exercise 1h

1 Given that set **P** = {2, 4, 6, 8} and set **Q** = {1, 2, 3, 4, 5, 6}, represent them on a Venn diagram and find;   (a) $n(P \cup Q)$ (b) $n(P \cap Q)$

2 Given that set R = { ⬜, ☐, ▱, ▱, ▱ }, and

   set S = { ◇, ◺, ☐, ▱, △, ▷ }

Show on Venn diagrams, the number of elements for R and S, hence find

(i) n(R ∩ S)   (ii) n(R ∪ S)   (iii) n(R ∩ S)P   (iv) n[(R ∩ S)) ∪ (RP ∩ S)]

3 Given the Venn diagram that follows,

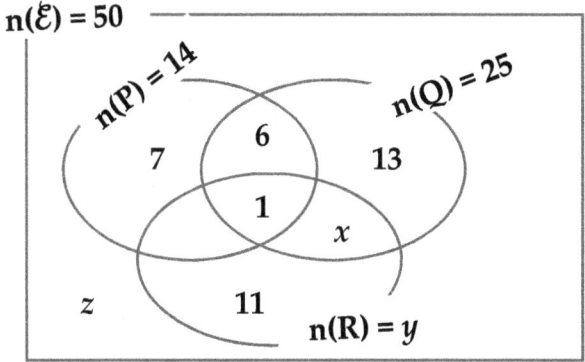

n(ℰ) = 50

n(P) = 14

n(Q) = 25

7   6   13

1

x

z   11

n(R) = y

(a) Find the values of letters x, y and z

(b) Find   (i) n(P ∩ Q)
            (ii) n(Q ∪ R)

1.7.3 Problems involving Venn diagrams

The concept of representing sets on Venn diagrams can also be extended to solving problems involving sets. These can be elementary or applied (that is, in daily life), tackled using word problems and illustrations.

Example 2

In a class of 48 students, 29 like Matooke (M) and 30 like Posho (P). Find the number of students who like both Matooke and Posho using a Venn diagram.

Solution

Let the number of those who like both Matooke and Posho be x
If the number of all students who like Matooke is 29, then those who like
Matooke only will be (29 − x). Likewise for Posho only is (30 − x).

n(ℰ) = 48

n(M) = 29   n(P) = 30

(29 − x)   x   (30 − x)
(29 − 11)   11   (30 − 11)

18   19

But; n(M only) + n(M ∩ P) + n(P only) = n(ℰ)

$$(29 - x) + x + (30 - x) = 48$$
$$29 - \cancel{x} + \cancel{x} + 30 - x = 48$$
$$59 - x = 48$$
$$^-x = 48 - 59$$
$$\frac{^-\cancel{1}x}{^-\cancel{1}} = \frac{^-11}{^-1}$$
$$\Rightarrow x = 11$$

Therefore, there are 11 students who like both Matooke and Posho.

Check: *Is 48 = 18 + 11 + 19 ?   Yes.*
            So, our calculations to find the value of x are accurate.

Example 2

In a class of 82 students, 50 like Mathematics (M) and 44 students like English (E). 4 students like neither of the two subjects. Find the number of students who like both Math and English.

Solution

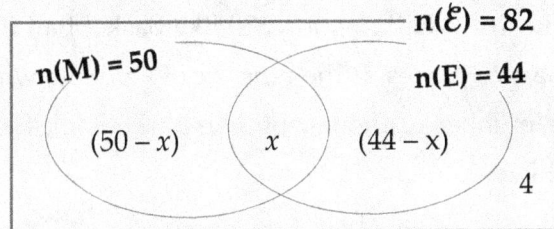

$n(\mathcal{E}) = 82$

$n(M) = 50$

$n(E) = 44$

$(50 - x)$ x $(44 - x)$

4

$n(\xi)$ $= 82$
$n(E)$ $= 44$
$n(M)$ $= 50$
$n(M \cup E)'$ $= 4$

$(50 - x) + x + (44 - x) + 4 = 82$
$50 - x + x + 44 - x + 4 = 82$
$98 \quad - \quad x \quad = 82$
$ \bar{\ }x \quad = 82 - 98$
$ \bar{\ }x \quad = \bar{\ }16$
$\Rightarrow x \quad = 16$

Therefore, there are 16 students who like both Math and English in the class.

Exercise 1*i*

1   In a class of 48 students 20 like beans (B) and 30 like peas (P). How many students like both beans and peas?

2   In a class of 50 students 25 like meat (M) and 5 like both meat and chicken. How many students like chicken only?

3   Of the 86 people who went for a film show 52 liked a Chinese (C) film only and 20 liked a Japanese (J) film only. How many people liked both the Japanese and Chinese films?

4   Given the Venn diagram next page:
    Find the value of x and hence find

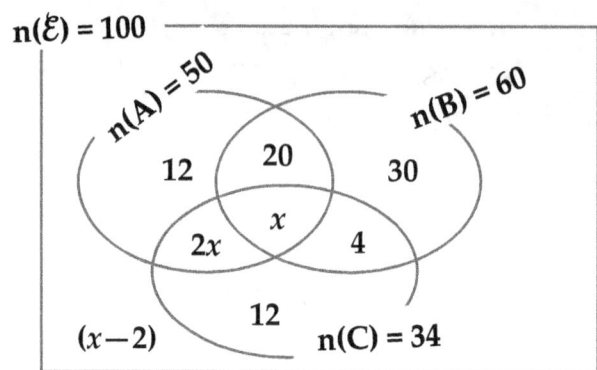

(i) n(A ∩ B ∩ C)
(ii) n(A ∩ C)
(iii) n(A ∪ B ∪ C)'
(iv) n(A ∩ B)

5   In a class of 58 students, 20 like volleyball (V) only, 30 like Basket ball (B) and the rest like neither of the two games. If the number of students who like both the games is ⅓ the number of those who like basket ball, find the number of students who like:

(i) Basket ball only?   (iii) Volley ball?

(ii) Both of the games?   (iv) Neither of the two games?

REVISION EXERCISE 1

1 (a) Name the following sets

(i)   **S** = {pens, pencils, punching machine, clips, ruler, markers }

(ii) **E** = {calculator, mobile phone hand set, digital watch, computer}

(iii) **C** = {4, 6, 8, 9, 10, 12, 14, 15 …}

(iv) **B** = {Chicken, Duck, Turkey, Pigeon}

(v)   **S** = {Biology, Physics, Chemistry}

(vi) **R** = {Lizard, Snake, Chameleon}

(vii) **G** = {Closed plane figures with all sides and angles congruent}

(viii) **L** = {Chord, Center, Diameter, Arc, Sector}

(b) List down the members of the following sets:

(i)   **M** = {Even multiples of 7 that are divisible by 6}

(ii) **S** = {Square numbers that are cubic and even but less than 100}

(iii) **N** = {All letters forming your surname}

(iv) **T** = {Multiples of ten that are divisible by 6}

(v)   **I** = {All instruments found in a geometry set}

(vi) **E** = {Multiples of 7 which are prime}

2 Given the sets below;

   V = { All vowels of the English alphabet},

   E = { All letters of the English alphabet},

   P = { 2, 3, 5, 7, 11, 13} ,   **N** = { A, B, C, D, E, F},

   Q = { All even numbers which are square numbers},

   R = {1, 3, 6, 10, 15 ...} and **C** = { 4, 6, 8, 9, 10, 12, 14, 15 ...}

   (a) State *true* or *false*

   (i) $V \subset E$ (ii) $N \equiv P$ (iii) **V** = {*a, e, i, o, u, w*}

   (iv) **R** = {Triangular numbers}   (v) n(E) = 27   (vi) n(Q) = 1

   (vii) $P \cap C = \varnothing$ (viii) $21 \in C$

   (b) State which of the given sets are finite and infinite

3   (a) Given the set D = {All consonants in the word "*diagrammatically*"}. Find
       the number of subsets which can be obtained from set D.

   (b) Given the sets: **A** = { 1, 2, 3, 4 ... 10} , **B** = { 0, 2, 4, 6 ... 10}

                        C = { 1, 4, 9 ... 100}   , **D** = { 1, 8, 27 ... 1000} , Find

   (i) n(D)   (ii) $C \cap D$ (iii) $(A \cup B) \cap C$ (iv) n(A $\cap$ B)   (v) A $\cap$ D

4

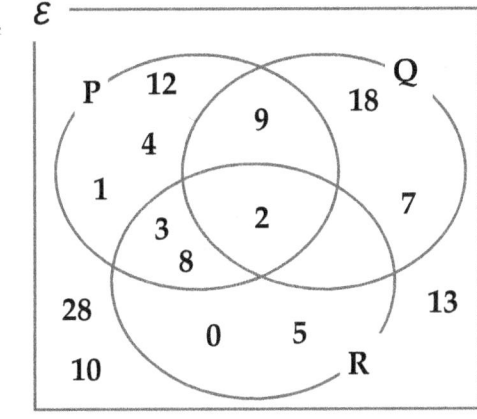

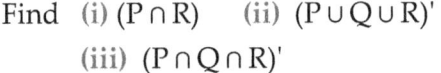

5

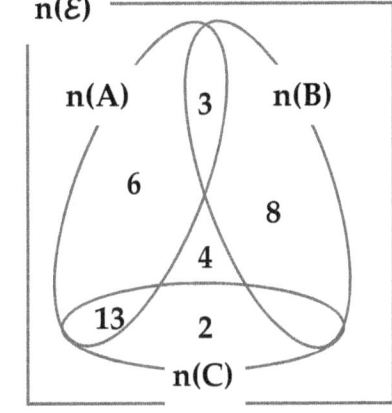

Find (i) (P $\cap$ R)   (ii) (P $\cup$ Q $\cup$ R)'   Find (i) n(A $\cup$ B $\cup$ C)' (ii) n(B $\cap$ C)
      (iii) (P $\cap$ Q $\cap$ R)'   (iii) n(A $\cap$ B $\cap$ C) (iv) n(A)

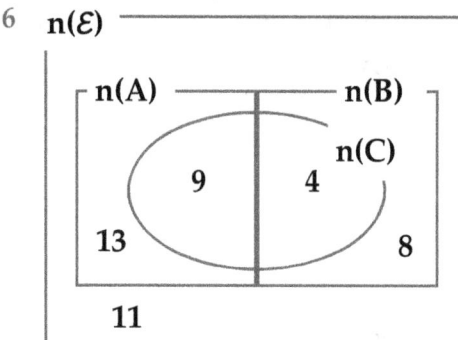

6 n(ℰ)

n(A) ———————— n(B)

n(C)

9   4

13   8

11

Find   (i) n(A ∪ B)′
       (ii) n(C)
       (iii) n(B ∩ C)
       (iv) n(A ∩ B)

7 n(ℰ) = 48

n(A) ———————— n(B)

13

11   2x

4   n(C)

(x + 5)

Find   (i) x
       (ii) n(B ∩ C)
       (iii) n(A ∪ B ∪ C)′
       (iv) n(C)

8   In a group of 82 students, 47 like Volleyball, 32 like Basket ball and 15 like neither of the two games.

   (a) Draw a venn-diagram to represent this information.

   (b) How many students like both games?

   (c) How many students like only one game?

9   In a class of 84 students, 30 like only Maths, 42 like only English. The number of those who like neither of the two subjects is twice the number of those who like both of the subjects.

   (a) Draw a Venn- diagram to represent this information.

   (b) How many students like both the subjects?

   (c) How many students like neither of the two subjects?

10   In a group of 52 students, 30 like Physics, 24 like Chemistry and twice the number of those who like both the subjects exceeds the number of students who like neither of the two subjects by 5.

   (a) Draw a Venn-diagram to represent this information.

   (b) How many students like both the subjects?

   (c) How many students like neither of the subjects?

2

NUMBER THEORY AND NUMERATION

INTRODUCTION

During ancient times people learnt how to count and measure; however there was a difficulty in recording the outcomes. It was therefore necessary to invent a way of taking and keeping records as well as computing. Piles of sticks or stones were first used to keep records, then *"tally sticks"*, which were the earliest form of keeping written records followed.

2.1 Numeration systems

A numeral is a symbol used to represent a number. If we have a number, twenty-four, we can normally write it as 24. However; using other numeration systems, we can write it differently, e.g., it is written as XXIV using Roman numerals.

Several systems of writing numerals have existed from the ancient to modern times. The Egyptian, Roman, Greek and Babylonian were among the ancient systems of numeration. These systems were based on additive, subtractive or place value ideas using pictures, marks or alphabet as symbols to represent numerals.

2.1.1 Egyptian number system

The Ancient Egyptians used an additive number system with the symbols as I for 1, ∩ for 10, ♀ for 100, ✍ for 1000, which were arranged starting from the right towards the left while forming a numeral.

For example,   I I ∩ ∩
              I I ∩ ∩ ♀ ✍   ≡   4 + 40 + 100 + 1000   = 1144

Activity 2.1

(a) Write in Egyptian number system: (i) 23   (ii) 315   (iii) 1048

(b) Express the following to numbers we use: (i) I I ∩   (ii) I I ∩ ∩ ♀
                                                        I ∩ ∩ ∩ ♀

2.1.2 Early Roman number system

This was an additive number system with the basic symbols as $I \equiv 1$, $V \equiv 5$, $X \equiv 10$, $L \equiv 50$, $C \equiv 100$, $D \equiv 500$, $M \equiv 1000$.

For example;

$$VIII \equiv 5 + 1 + 1 + 1 = 8,$$
$$XXX \equiv 10 + 10 + 10 = 30,$$
$$CCXV \equiv 100 + 100 + 10 + 5 = 215, \text{ and}$$
$$LXXXXVI \equiv 50 + 10 + 10 + 10 + 10 + 5 + 1 = 96.$$

Activity 2.2
(a) Write down the meaning of (i) CXXXII (ii) CCCCLXXVI
(b) Express the following as Roman numerals (i) 123 (ii) 519

2.1.3 Later Roman number system

This extends the Early Roman system to the subtractive idea, and this saves space. The symbols had to be written in order of size, starting from left towards the right;

- If a *larger symbol* comes first, then it means *addition*;
- And if the *smaller symbol* comes first, then it means *subtraction*.

This technique only works for two adjacent symbols.

For example; LXV means: $50 + 10 + 5$ $= 65$ (*Additive*)

IX means: $(X - I)$ or $10 - 1$ $= 9$ (*Subtractive*)

DVC means: $D + VC = D + (C - V)$

$500 + (100 - 5)$ $= 595$

Activity 2.3

Compare the following number systems by completing the following table.

	Number	Early Roman System	Later Roman System
1	580	DLXXX	DLXXX
2	93	LXXXXIII	XCIII
3	198	_____	CXCVIII
4	1989	_____	_____
5	_____	DCCCLXXXXVIIII	_____

2. Number Theory and Numeration

Remember that we wrote 96 as LXXXXVI in the Early Roman number system, but under the Later Roman number system , it is written as;

XCVI meaning: $(100 - 10) + (5 + 1)$ $= (90 + 6)$ $= 96$

2.1.4 Hindu-Arabic Number System

The ancient systems of numeration were based on additive, subtractive (for Later Roman) or Place value (for Babylonian). The idea of using pictures, marks or letters for symbols to represent numbers was also used. These were very cumbersome to handle due to the large groups of symbols used to represent a particular numeral. And in most of them it was hard to go beyond thousands.

Eventually the Hindu-Arabic number system was introduced as a modern system and it is the one we use. It is based on 10 digits as symbols and place value. It is better than all the other previous systems of numeration.

2.2   Uses of numbers

You have earlier studied numbers, their four basic operations (addition, subtraction, multiplication and division); and some number applications.

You could have realized that numbers are used;

1 For counting, measuring and comparing quantities

2 To label and identify items. For example, numbers are used to label items like TV channels, school grades or book pages and so on. Codes such as telephone numbers, ID numbers, post box numbers, etc are used in identification of items.

3 To represent situations according to preference, for example in football a win is denoted 3 points, a draw 1 point and a loss is denoted no point (or 0 point).

4 To show where some thing is in a reference system. For example 37°C represents the normal human body temperature on the Celsius scale. The time 9:25 AM is time in the morning based on a 12 hour clock system.

We also use number applications through calculations to find unknown values of things, to find how values relate and so on.

Activity 2.4

1 If I asked you, *"in which class are you?"* You would answer, *"I am in senior 1."* Discuss other situations or items where you can use numbers and state how they are used.

2 A TV is showing channel 12 and I want to view channel 40. If the TV has 55 channels in total.
   (a) Do I need to scroll forward or backwards in order to reach my favorite channel faster?
   (b) How many channels do I have to bypass in order to reach it faster?

3 Explain how numbers apply to the following situations;
   (i) Calendars   (iii) Piece of land
   (ii) A can of water   (iv) Radio station frequencies

2.3   Kinds of numbers

Study carefully the different kinds of numbers in figure 2.1 next page, and ensure that;

   1 You understand the characteristics of each number category.

   2 You are able to differentiate between any two given number categories.

On the illustration the different kinds of numbers are each enclosed separately, though within another set enclosure.

   • All Natural (or Counting) numbers are Whole, Integers, Rational and Real.

   • All Whole numbers are Integers, Rational and Real.

   • All Integers are Rational and Real.

   • All Rational numbers are Real

   • All Irrational numbers are Real

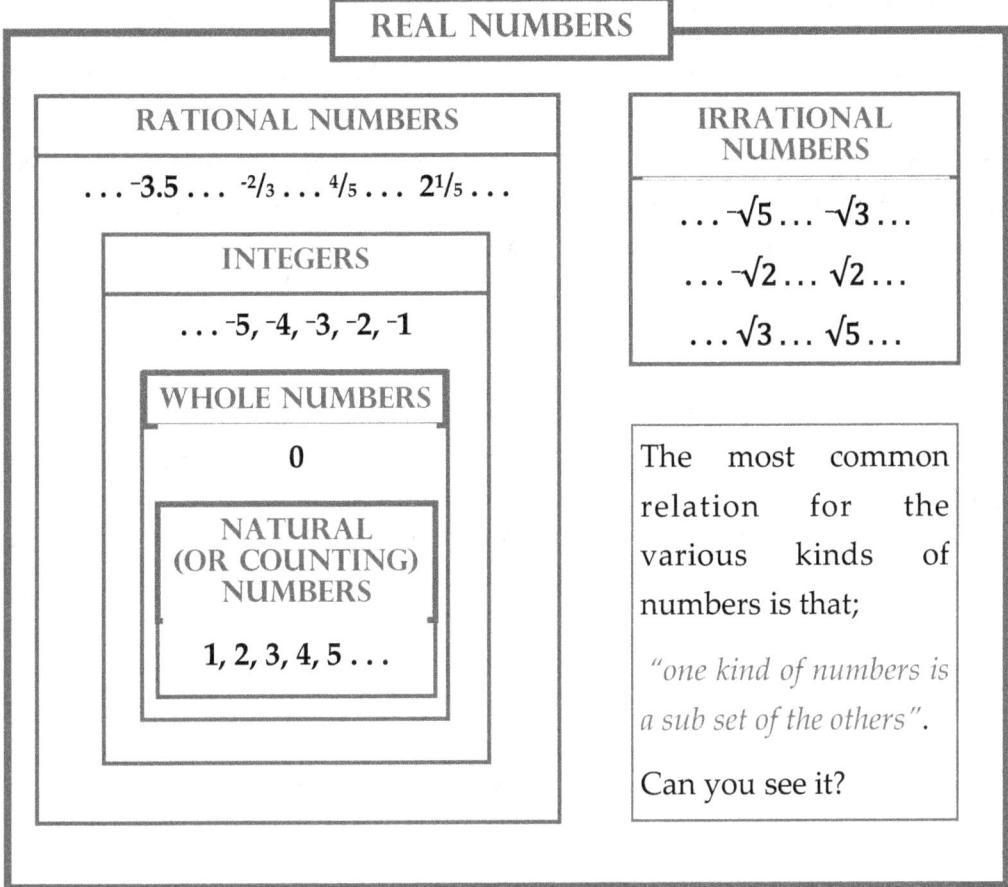

REAL NUMBERS

RATIONAL NUMBERS

... ⁻3.5 ... ⁻²/₃ ... ⁴/₅ ... 2¹/₅ ...

INTEGERS

... ⁻5, ⁻4, ⁻3, ⁻2, ⁻1

WHOLE NUMBERS

0

NATURAL
(OR COUNTING)
NUMBERS

1, 2, 3, 4, 5 ...

IRRATIONAL
NUMBERS

... ⁻$\sqrt{5}$ ... ⁻$\sqrt{3}$ ...

... ⁻$\sqrt{2}$ ... $\sqrt{2}$ ...

... $\sqrt{3}$ ... $\sqrt{5}$ ...

The most common relation for the various kinds of numbers is that;

"one kind of numbers is a sub set of the others".

Can you see it?

Fig. 2.1

2.3.1 Real numbers

This is a set of numbers constituting of both rational and irrational numbers. From the illustration, the set of real numbers given is;

$\mathbb{R} = \{\ldots ⁻5, ⁻4, ⁻3.5, ⁻3, ⁻\sqrt{5}, ⁻2, ⁻\sqrt{3}, ⁻\sqrt{2}, ⁻1, ⁻²/₃, 0, ⁴/₅, 1, \sqrt{2}, \sqrt{3}, 2, \sqrt{5}, 3, 4, 5 \ldots\}$

2.3.2 Rational numbers

These are numbers which can be expressed in the form $^a/_b$, where $b \neq 0$ and both a and b are integers.

From the illustration, the set of rational numbers given is:

$\mathbb{Q} = \{\ldots ⁻5, ⁻4, ⁻3.5, ⁻3, ⁻2, ⁻1, ⁻²/₃, 0, ⁴/₅, 1, 2, 3, 4, 5 \ldots\}$

2.3.3 Irrational numbers

These can only be written in a numeric form as never-ending and non-repeating decimal fractions. They can not be written in the form $^a/_b$ as for rational numbers. The square root of any prime number is irrational. π is also irrational, written as a decimal which never ends, but with no repetitive digits like those for recurring decimals. From the illustration, the set of irrational numbers is given as; $\mathbf{I} = \{\ldots -\sqrt{5}, -\sqrt{3}, -\sqrt{2}, \sqrt{2}, \sqrt{3}, \sqrt{5} \ldots\}$

2.3.4 Integers

These include zero, positive and negative numbers. Positive integers are also known to be natural numbers. Integers are also known as signed or directed numbers. This is due to the fact that numbers are assigned positive or negative signs to reveal their direction. From the illustration, the set of integers is; $\mathbb{Z} = \{\ldots -5, -4, -3, -2, -1, 0, 1, 2, 3, 4, 5 \ldots\}$

2.3.5 Whole numbers

These include zero and positive integers. They are non-fractional. From the illustration, the set of whole numbers; $\mathbf{W} = \{0, 1, 2, 3, 4, 5 \ldots\}$

2.3.6 Natural numbers

This is a set of numbers 1, 2, 3, 4, 5 and so on used for counting. They are also known as counting numbers. From the illustration, the set of natural numbers; $\mathbb{N} = \{1, 2, 3, 4, 5 \ldots\}$

Activity 2.5

1 Differentiate between: (a) Integers and whole numbers

                                     (b) Whole numbers and natural numbers

2 State *true* or *false* in the following:

   (a) Real numbers are either rational or irrational numbers

   (b) All integers are whole numbers

   (c) All natural numbers are integers

   (d) 0 is neither whole nor natural

2.4 Whole numbers and Place value

The decimal (base ten) system of numerals has ten digits, that is 1, 2, 3, 4, 5, 6, 7, 8, 9 and 0. A single digit or a group of digits form a numeral and in this numeral each digit has a place value. In the number 2222, each of the four 2s has its value dictated by its place in the numeral.

2.4.1 Place values

In the number 2222, the first digit 2 is in the place of thousands, the second digit 2 is in place of hundreds, the third digit 2 is in the place of tens and the forth digit 2 is in the place of ones

$$
\begin{array}{rcl}
2\,0\,0\,0 & \Rightarrow & 2\text{ Thousands} \\
2\,0\,0 & \Rightarrow & 2\text{ Hundreds} \\
2\,0 & \Rightarrow & 2\text{ Tens} \\
+\quad 2 & \Rightarrow & 2\text{ Ones} \\
\end{array}
$$

Number: 2222 (*Two thousand two hundred twenty-two*)

Similarly, the digits 2, 5 and 9 can form six different numerals with help of place values. By changing the digitsNarrangement variously, the formed numerals (in ascending order) are:

(i)	(ii)	(iii)	(iv)	(v)	(vi)
H T O	**H T O**	**H T O**	**H T O**	**H T O**	**H T O**
2 5 9	2 9 5	5 2 9	5 9 2	9 2 5	9 5 2

The place value chart that follows will guide you on how place values are named. Using the numbers (i) 813152378021 and (ii) 7803921182 we are going to go see how place values help us identify numbers.

	HB	TB	B	HM	TM	M	HTh	TTh	Th	H	T	O
(i)	8	1	3	1	5	2	3	7	8	0	2	1
(ii)			7	8	0	3	9	2	1	1	8	2

Fig. 2.2

Remember, that the following abbreviations will help us easily represent the place values.

Hundreds of Billions (HB)	Tens of Billions (TB)	Billions (B)
Hundreds of Millions (HM)	Tens of Millions (TM)	Millions (M)
Hundreds of Thousands (HTh)	Tens of Thousands (TTh)	Thousands (Th)
		Tens (T)
		Ones (O)

Activity 2.6

From the place value chart in fig. 2.2 on page 30;

(a) What is the place value of 5 in the number in part (i)?

(b) How many hundreds of thousands are in the number in parts (i) and (ii)?

(c) Write in words the numbers in parts (i) and (ii).

2.4.2 Writing numbers using place values

You could have realized in activity 2.6 that writing in words is difficult in the way the place values are as shown before in the chart (*fig.2.2*). In easier arrangements, we have the place values subdivided into groups of three digits starting from the right towards the left hand side.

	BILLIONS			MILLIONS			THOUSANDS			UNITS		
	H	T	O	H	T	O	H	T	O	H	T	O
(i)	8	1	3	1	5	2	3	7	8	0	2	1
(ii)			7	8	0	3	9	2	1	1	8	2

Fig. 2.3

In words we have;

    (i)   813 Billions, 152 Millions, 378 Thousands, 021; which gives;

Eight hundred thirteen *billion* one hundred fifty-two *million* three hundred seventy-eight *thousand* and twenty-one.

(ii) Alternatively make groups of threes from the right to the left, separated with commas.

7, 803, 921, 182 giving;

7	803	921	182
↓	↓	↓	↓
Billions	Millions	Thousands	Units

Seven billion eight hundred three million nine hundred twenty one thousand one hundred eighty-two.

Given a number in words to write it in figures, say; Eleven billion two hundred eighty-eight million three hundred forty-two thousand one hundred twelve.

We have:-

Eleven billion ⟶ **11,000,000,000**

Two hundred eighty eight million ⟶ **288**,000,000

Three hundred forty two thousand ⟶ **342**,000

One hundred twelve ⟶ + **112**

11,288,342,112

Hence: **11,288,342,112** in figures.

Activity 2.7

(a) Write in words: (i)   100021   (ii)   3007002

(b) Write in figures:
- (i) Two hundred nine millions six hundred forty two thousand nine hundred eighty-two.
- (ii) Forty-two thousand eight.

Exercise 2a

---

1 Write in words the following figures

   (i)   3642   (iv)   8986764012   (vii) 64570001301

   (ii) 41138   (v)   3042681243   (viii) 818110392830

   (iii) 148390981   (vi) 30002138261

2 Write the following as figures
   (i)   Forty two thousand nine hundred twenty three
   (ii)   Four thousand sixty five
   (iii) Two thousand one hundred fifteen
   (iv) Seven billion one hundred thirty two million two hundred one thousand one hundred fifty three
   (v)   Three hundred twenty nine billion five hundred three thousand twelve.
   (vi) Nine hundred twelve billion three hundred thirty-one.

3 What is the place value of 3 in the figures that follow?
   (i) 3000484   (ii) 842311   (iii) 3482914
   (iv) 864311201   (v) 13000413516

4 Given the digits 4,6,9,5 and 3
   (a) Write in figures;
      (i) The smallest number that can be obtained from these digits, writing each digit once.
      (ii) The largest number that can be obtained from these digits, writing each digit once.
   (b) What is the place value of 4 in; (i) the number obtained in a(i) above
                                   (ii) the number obtained in a(ii) above.

2.5   Revision on whole number operations

2.5.1 Addition
Example 1

Add: (i) 349207 + 831802   (ii) 20999 + 88891 + 3602

   Solution

(i)
```
   1  1
   3 4 9 2 0 7
 + 8 3 1 8 0 2
 1 1 8 1 0 0 9
```

(ii)
```
    1   2   1   1
    2 0 9 9 9
    8 8 8 9 1
 +    3 6 0 2
  1 1 3 4 9 2
```

Activity 2.8

Fill in the missing digits

(i)
```
    1 □ 4 □
  +  4 3 □ 6
    □ 7 3 8
```

(ii)
```
      4 8 □ 4 □
      □ 1 1 □ 2
  +  1 2 □ 4 1
      6 □ 0 9 1
```

Exercise 2b

Add the following numbers

1.
```
   2 3 0 4 5 6
 +     3 4 3 2 1
```

2.
```
   6 3 1 0 3
   1 0 1 3
 + 5 4 8 2
```

3.
```
     3 4 9 2 0 7
 + 8 3 1 8 0 2
```

4. 4321 + 1420

5. 1230 + 16001

6. 242814 + 3849

7. 241cm + 3498cm + 33001cm

8. 342g + 3082g + 4261g + 7029g

Filling in the missing values

9.
```
   *  4 1 *
 + 5 8 *  4
   9 *  2 8
```

10.
```
   *  9 *  3 3
 + 2 4 2 4 *
   9 4 *  7 9
```

11. In a school there were 890 male students and 362 females, how many students has the school got altogether?

12. A book has 906484 words and the other has 240062 words. How many words are there in the two books altogether?

13. When three boys went to a farm to collect oranges for sale, one picked 42482 fruits, the second picked 8248 fruits and the third picked 60287. How many fruits were picked by the three boys altogether?

14. Below are the litres of milk sold by Ali in 4 months.

April	May	June	July
8,008	20,428	6,242	10,628

How many litres of milk did he sell altogether in the four months?

15   A school was grouped into three such that students go for general cleaning work.

Group A	Group B	Group C
385	664	754

How many students went for general cleaning altogether?

2.5.2   Subtraction

Example 2

Subtract: 45334 – 2040 – 3418

Solution

```
  4 5 3 3 4
−     2 0 4 0
  4 3 2 9 4  → (Result)
−     3 4 1 8.
  3 9 8 7 6  → (Final result)
```

Deal with two figures at a time, each time carrying out subtraction as usually done. Always remember to borrow, where necessary

---

Activity 2.9

Fill in the missing digits

(i)
```
    4  5 □ 4 □
  − □  1 0 2 1
    1 □ 5 □ 6
```

(ii)
```
    8 □ 4 □
  −   2 5 2
  □ 0 □ 4
```

Example 2

A library caught fire, but 3 shelves with 4123, 14821 and 4200 books remained undestroyed. How many books were destroyed if the library had 828462 books altogether?

Solution

```
  8 2 8 4 6 2
−     4 1 2 3
  8 2 4 3 3 9
−   1 4 8 2 1
  8 0 9 5 1 8
−     4 2 0 0
  8 0 5 3 1 8
```

Subtract the number of the books undestroyed from the total number of books, which were in the library.

∴ 805318 books were destroyed by fire

Exercise 2c

Subtract: 1 81024 – 2410   2 21102 – 8984

              3 42035 – 4624 – 3004   4 911204mm – 20656mm

              5 4660312Kgs – 998209Kgs

6 A tank containing 98,600 litres of water was damaged and 2836 litres were lost. How many litres remained in the tank?

7 A driver always covers 54,110,645 metres in 15 hours but when he fell sick he covered 3,821,424 metres in 15 hours. How many metres did he fail to cover due to sickness?

8 In a year of 365 days, Dan did not work in his garden for 124 days. How many days did he go to work in his garden?

9 1258 cars passed a certain point of the road in 1 hour yesterday and today 3281 cars passed that same point in 1 hour. What is the difference in the number of cars passing that point today and yesterday?

10 In a crowd of 72,382 people listening to the presidential candidatesN speeches, 4,841 felt disgusted and left the rally. How many people remained listening to the candidates?

2.5.3 Multiplication

Example 1

Multiply: 6082 x 24

Solution

```
        6  0  8  2
    x         2  4
    2  4  3  2  8 ……………………(6082 x 4)
+ 1  2  1  6  4  0 …………………(6082 x 20)
  1  4  5  9  6  8 …………………(6082 x 24)   ← Final answer
```

Activity 2.10

Fill in the missing digits:

```
              ☐  1  ☐  2
    x               3  4
        1  6  5  2  8
  + 1  2  3  ☐  6  0
    1  3  0  4  8  ☐
```

Exercise 2d

Multiply:-   1 331 x 28   2 13 x 33 x 44   3 2341 x 88

4 841 x 232   5 32470 x 420

6 In a video hall 1238 people paid $25 each for a film show. How much money was collected by the video hall authorities?

7 420 students in a school misbehaved and they were punished by paying a fine of £121 each to rehabilitate the premises they demolished. How much money did the school collect altogether?

8 When painting, a painter uses 31 tins for one house. How many tins of paint does he need to paint 238 similar houses?

2.5.4 Division

Example 2

(a) Mr. Kayizzi uses 5600 litres of paint on 40 rooms. How many litres of paint are required to paint one room of the same kind?

(b) In a club there are 45 people who paid the entry fee of sh. 38250 altogether. How much did each person pay?

Solution

(a) 1 room requires (5600 ÷ 40) litres = 140 litres

(b) Each person paid (38250 ÷ 45) = shs.850 per head

Activity 2.11

(a) Work out the long division for 344974 ÷ 82

(b) Fill in the missing digits:

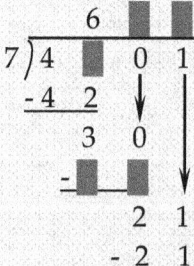

Exercise 2e

Divide: 1   282480 by 30   2   142800 by 400   3   3487 by 11

4   156520 by 43   5   91410 by 110

6 A water tank has 3824 litres of water, how many 8 litres container will I scoop from the tank to make it empty.

7 To make 483 tables I needed 1449 pieces of timber, how many pieces of timber do I need to make one table?

8 How many tiles of 400cm² area each do I need to cover an area of 130000cm² of floor?

2.5.5   Mixed operations

In questions with two or more operations the order of operation is "Brackets" then $\mathbb{N}$Of$\mathbb{N}\mathbb{N}$ivision$\mathbb{N}\mathbb{N}$ultiplication$\mathbb{N}\mathbb{N}$ddition$\mathbb{N}$and finally $\mathbb{N}$ubtraction.$\mathbb{N}$For a reminder, the short form, **BODMAS** is always used.

Example 1

Evaluate: $32 + (42 + 8) \times 25 - 43$

Solution

$$32 + \underbrace{(42 + 8)} \times 25 - 43$$

$$32 + \underbrace{50 \times 25} - 43$$

$$\underbrace{32 + 1250} - 43$$

$$1282 - 43$$

$$1239$$

Here, carry out the *brackets* first, followed by *multiplication*, *addition* and finally *subtraction*.

$\therefore 32 + (42 + 8) \times 25 - 43 = 1239$

Activity 2.12

I had 12 coins of sh. 200 each, 3 coins of sh.500 each and 13 paper notes of sh.1000 each.

(a) How much money did I have altogether?

(b) How many sh.50 coins can I obtain from this amount?

(c) What is the least number of coins I can obtain from this amount of money ? Give the denomination (s) of these coins.

(Note that the following coin denominations exist,sh.50, sh.100, sh.200 and sh.500)

Exercise 2f

Evaluate:

1 $(220 \div 11)$ of $4 + 450 - 30$ 2 $(124 - 108) + 32 \times 6$ of $42 - 33 \div 11$

3 $134 \times 408 \div 40 - (411 - 352)$ of 12 4 $(342 \div 3) + 42$ of $62 - 3 \times 8$

5 42×15 of $(80 \div 4) - 320 \div 4 + 640$

6 Four machines produce 324,306,640 and 520 pieces of bricks per hour.

  (i) How many bricks are produced in 1 hour by all the 4 machines altogether?

  (ii) If the machines work for 6 hours a day, how many pieces of bricks will be produced in 3 days by all the 4 machines?

  (iii) If the 4 machines work for 8 hours, and 34 and 42 bricks from two of the machines are rejected as damages per hour. How many bricks will be produced by the 4 machines in 8 hours?

7 At MakaN shop, Makula bought 42 books at sh.320 each, 240 pens at sh.150 each and 13 file folders each at sh. 1500. it was later discovered that 7 books, 32 pens and 1 file folder were faulty and these were not to be paid for. Find how much money Makula;

  (i) Had to pay originally?
  (ii) Paid after a deduction due to faulty items?

8 As Norah was arranging 34 glasses in a cupboard after washing them, she accidentally broke 5 of them. Her mother bought more 11 glasses and they were also put in the cupboard. When visitors came, Norah removed all the glasses from the cupboard and arranged 5 glasses on each table in preparation for the visitors.

  (a) (i) How many glasses were there altogether?

     (ii) How many tables were used to prepare for visitors?

  (b) It was later discovered that each table had two excess visitors.

     (i) How many visitors were they altogether?

     (ii) How many more glasses were bought to make sure that every visitor had a glass?

     (iii) How much was the cost of the more glasses bought?

2.6 Indices

Numbers can be written in a shorter form, where the idea of multiplying the same number upto a given number of times is used. Under such operations a number can be reduced to another representation by use of indices. Values such as 1000, 10000, 100000 etc, can be expressed as indices say;

1000	= 10 x 10 x 10 (*multiplying three 10s*)		$\Rightarrow 10^3$ (as an index)
10000	= 10 x 10 x 10 x 10 (*multiplying four 10s*)		$\Rightarrow 10^4$ (as an index)
100000	= 10 x 10 x 10 x 10 x 10 (*multiplying five 10s*)		$\Rightarrow 10^5$ (as an index)

A value, 10^2 means ten to the 2nd power, ten squared or multiply 10 by itself.

Similarly, 10^3 means ten to the 3rd power, ten cubed or multiply three 10s.

In 10^2 and 10^3, 2 and 3 are called the *exponents* or *powers* and 10 is the *base*. We can have powers to other bases say; 2^4, 3^2, 4^3, 11^4, etc.

2^4 can be read as *2 power 4 (or 2 exponent 4)*. That is, 2^4 — *Exponent or power or index*

⤷ *Base*

2.6.1 Expressions involving Indices

Have you realized that for the powers of 10 the number of zeros is the same as the exponent or power? Can you write the indices from the multiples of ten? We can write whole numbers as indices, for example 32 can be expressed as 2^5 and so on. Given a value 81, we can express it in terms of its prime factors and then later as indices.

Example 1

Express as indices, the following numbers: (i) 32   (ii) 81   (iii) 100

Solution

(i)	32	(ii)	81	(iii)	100
	2 x 16		9 x 9		2 x 50
	2 x 2 x 8		3 x 3 x 3 x 3		2 x 2 x 25
	2 x 2 x 2 x 4		$\therefore 81 = 3^4$		2 x 2 x 5 x 5
	2 x 2 x 2 x 2 x 2		(or $81 = 9^2$)		$\therefore 100 = 2^2 \times 5^2$
	$\therefore 32 = 2^5$				(or $100 = 10^2$)

Remember that any number to power 1 gives that same number, for example; $10^1 = 10$, $3^1 = 3$, $a^1 = a$, and so on.

Given indices, they can also be represented as numbers. If you are given an index written as: $10 \times 10 \times 10 \times 10 \times 10 \times 10 \times 10$ (or 10^7), can you write it as a natural number?

Example 2

Write the following as indices: (i) $23 \times 23 \times 23 \times 23$ (ii) $a \times a \times a \times a \times a$

Solution

(i) $23 \times 23 \times 23 \times 23 = 23^4$

(ii) $a \times a \times a \times a \times a = a^5$

Example 2

Write the following indices as factors: (i) 9^4 (ii) $13^2 \times 14^3$

Solution

(i) $9^4 = 9 \times 9 \times 9 \times 9$
   (or $3 \times 3 \times 3 \times 3 \times 3 \times 3 \times 3 \times 3$)

(ii) $13^2 \times 14^3 = 13 \times 13 \times 14 \times 14 \times 14$
    (or $2 \times 2 \times 2 \times 7 \times 7 \times 7 \times 13 \times 13$)

Note: Most students have a tendency of expanding, say 9^4 as 9×4. This is WRONG; $9^4 \neq 9 \times 4$. Generally always remember that; $a^b \neq a \times b$. It is known to be: $a^b = a \times a \times a \ \dots \ upto \ b \ times.$

Exercise 2g

Express the following numbers as indices

1 108	2 625	3 10,000	4 1080
5 40×20	6 120	7 1331	8 2048

Write the following factors as indices:

9   $2 \times 2 \times 2 \times 2 \times 2$

10   $9 \times 9 \times 9 \times 9 \times 9 \times 9 \times 9$

11   $14 \times 14 \times 14$

12   $5 \times 5 \times 5 \times 6 \times 7 \times 7$

13   $3 \times 3 \times 3 \times 9 \times 9 \times 9 \times 13 \times 13$

14   $25 \times 7 \times 7 \times 7 \times 32$

Write the following indices as factors:

15 11^3	16 28^4	17 10^6	18 111^3	19 2^4

Write the following indices as numbers:

20 10^4	21 9^3	22 3^5	23 8^3	24 11^4

2.6.2 Operations with Indices

Multiplication

To multiply indices, we shall use a general rule of operation.

To multiply indices with the same base, we add the exponents and maintain the base, that is,

$$a^x \times a^y = a^{(x+y)}$$

Example 1

Work out he following: (i) $2^3 \times 2^4$ (ii) $11^3 \times 11^1$

Solution

(i) $2^3 \times 2^4 = 2^{(3+4)}$ (iii) $11^3 \times 11^1 = 11^{(3+1)}$

    $= 2^7$ $= 11^4$

Example 2

Work out: (i) $12^0 \times 12^1 \times 12^2$ (ii) $3^{-2} \times 3^1 \times 3^4$ (iii) $32 \times 8 \times 2^{-4}$

Solution

(i) $12^0 \times 12^1 \times 12^2 = 12^{(0+1+2)}$ (ii) $3^{-2} \times 3^1 \times 3^4 = 3^{(-2+1+4)}$

$\qquad\qquad\qquad = 12^3$ $\qquad\qquad\qquad = 3^3$

(iii) To work out 32, 8 and 2^{-4} using indices first put all of them under the same base throughout: $32 = 2^5$ and $8 = 2^3$

So, $32 \times 8 \times 2^{-4} = 2^5 \times 2^3 \times 2^{-4}$

$\qquad\qquad\qquad = 2^{(5+3+-4)}$

$\qquad\qquad\qquad = 2^4$

Division

To divide indices, we shall use a general rule of operation.

To divide indices with the same base, we subtract the exponent of the divisor from that of the dividend and maintain the base, that is,

$$a^x \div a^y = a^{(x-y)}$$

Mathematics Power

Example 1

Work out the following: (i) $3^4 \div 3^2$ (ii) $4^2 \div 4^5$ (iii) $11^1 \div 11^{-3}$

Solution

(i) $3^4 \div 3^2 = 3^{(4-2)}$
$\qquad\quad = 3^2$

(ii) $4^2 \div 4^5 = 4^{(2-5)}$
$\qquad\qquad = 4^{-3}$

(iii) $11^1 \div 11^{-3} = 11^{(1--3)}$
$\qquad\qquad\qquad = 11^4$

Example 2

Work out: $\dfrac{2^3 \times 5^{-2} \times 7^1 \times 7^{-2}}{2^1 \times 5^{-3} \times 7^2}$

Solution

$$\frac{2^3 \times 5^{-2} \times 7^1 \times 7^{-2}}{2^1 \times 5^{-3} \times 7^2} = \frac{2^3 \times 5^{-2} \times 7^{(1+-2)}}{2^1 \times 5^{-3} \times 7^2}$$

$$= \frac{2^3}{2^1} \times \frac{5^{-2}}{5^{-3}} \times \frac{7^{-1}}{7^2}$$

$$= (2^3 \div 2^1) \times (5^{-2} \div 5^{-3}) \times (7^{-1} \div 7^2)$$

$$= [2^{(3-1)}] \times [5^{(-2--3)}] \times [7^{(-1-2)}]$$

$$= 2^2 \times 5^1 \times 7^{-3}$$

Activity 2.13

1 Use indices to work out: (i) $3^{-1} \times 3^0 \times 3^1$ (ii) $125 \times \dfrac{1}{625} \times 5^2$

2 Use indices to work out: (i) $7^{-2} \div 7^5$ (ii) $\dfrac{3^2 \times 3^{-1} \times 5^2}{3^{-3} \times 5^{-1} \times 5^3}$

- *Be careful in cases where you have to deal with both multiplication and division in the same problem.*
- *Remember how to handle positives and negatives in a particular problem.*

Numbers to the power zero

Given to divide 32 by 32 we have; $32 \div 32 = 1$ ----------------- (i)

But $32 = 2^5$, expressed as an index is: $32 \div 32 = 2^5 \div 2^5$

$32 \div 32 = 2^{(5-5)}$

$32 \div 32 = 2^0$ --------------- (ii)

From eqn (i) $32 \div 32 = 1$ and from eqn (ii) $32 \div 32 = 2^0$, therefore; $2^0 = 1$.

Generally, any number except zero to power zero is 1. So, $n^0 = 1$, where $n \neq 0$.

Negative indices

We have indices with negative exponents say $2^{-1}, 3^{-2}, 10^{-1}$, etc. These negative indices can be expressed as positive indices, which are fractional in nature, that is,

$$3^{-2} = 1/3^2 = 1/9 \, ; \qquad 10^{-2} = 1/10^2 = 1/100 \, ; \quad 10^{-3} = 1/10^3 = 1/1000$$

Example 1

Express the following indices as positives: (i) 11^{-3} (ii) $1/2^{-3}$

Solution

(i) $11^{-3} = \dfrac{1}{11^3} = \dfrac{1}{11 \times 11 \times 11} = \dfrac{1}{1331}$

(ii) $\dfrac{1}{2^{-3}} = 2^3 = (2 \times 2 \times 2) = 8$

Example 2

Express the following indices as negatives: (i) $\dfrac{1}{5^2}$ (ii) $\dfrac{2}{3^2}$ (iii) $\dfrac{1}{2^2 \times 3^4}$

Solution

(i) $\dfrac{1}{5^2} = 5^{-2}$ (ii) $\dfrac{2^5}{3^2} = 2^5 \times 3^{-2} = \dfrac{3^{-2}}{2^{-5}}$ (iii) $\dfrac{1}{2^2 \times 3^4} = 2^{-2} \times 3^{-4}$

Always remember that if a number has a negative index, to change it to positive, we take its reciprocal but with an index with an opposite exponent.

Exercise 2h

Express the following indices as positive:

1 9^{-2} 2 3^{-5} 3 $\dfrac{2^{-2}}{3^{-1}}$ 4 $3^{-6} \times 2^3$ 5 $\dfrac{13^{-4} \times 6^{-2}}{3}$

Express the following as negative indices:

6 3^5 7 $\dfrac{1}{11^3}$ 8 $\dfrac{2^4}{8^3}$ 9 $\dfrac{3^2 \times 2^5}{6^{-3}}$ 10 $\dfrac{2^{-2} \times 3^3}{5^{-2}}$

Work out the following

11 $3^{-1} \times 3^2 \times 3^6$ 12 $4^3 \times 4^{-2} \times 4^2$ 13 $9 \times 9^3 \times 9^{-1}$

14 $5^{-2} \times 5^2 \times 5^0$ 15 $3^6 \times 3^{-7} \times 3^1$ 16 $5^8 \div 5^6$

17 $3^4 \div 3^1$ 18 $4^3 \div 4^{-2}$ 19 $3^{-2} \div 3^2$

20 $6^2 \div 6^3$ 21 $3^3 \div 9 \times 3^{-1}$

22 $\dfrac{2^4 \times 2^{-1} \times 2^0}{4}$ 23 $\dfrac{5^2 \times 5^{-6} \times 5^1}{5^3 \times 5^{-1}}$ 24 $\dfrac{4 \times 2^2 \times 2^{-1}}{4 \times 2^{-6}}$

25 $\dfrac{5^2 \times 5^{-3} \times 7^1 \times 11^{-2}}{5^3 \times 7^{-2} \times 11^3}$

REVISION EXERCISE 2

1 (a) Write in words the following figures   (i) 4300041   (ii) 111008211.
  (b) Write in figures the following:
      (i) Four hundred twenty- nine thousands.
      (ii) Four hundred twelve millions twenty- six.

2 Work out the following: (i) 31041 + 8012 + 381   (ii) 30801 +10003 – 8089
                                (iii) 342 x 401   (iv) 498135 ÷ 11 – 624

3 A crate of soda carries twenty-four bottles:
   (a) How many crates of soda do I need to serve 1452 visitors such that each
       of them gets 4 bottles of soda?
   (b) If each crate from the factory costs 9,600/=, how much money did I need
       to buy sodas for my visitors?
   (c) If I had bought sodas from the shop each bottle would cost me 500/=.
       Find how much money I save by buying in crates.

4 Work out the following;

   (i) $5^2 \times 5^{-1}$ (ii) $3^6 \div 3^7$ (iii) $\dfrac{2 \times 3^4 \times 5^0}{3^2 \times 2^{-1} \times 5}$ (iv) $\dfrac{(3^4 \div 3^{-1}) \times 3^{-2}}{100^0 \times 3^3}$

3 NUMBER BASES

INTRODUCTION

When counting, some times we use dozens (groups of 12), and for tallies in statistics we tie 5 tally sticks for a bundle. Remember the bundles of sticks which were grouped in 10s while you were still in lower primary school. We normally count in lots of ten. Counting in lots of ten is the same as working in base ten or *denary* or *decimal* base. Similarly if we count in twos, it means we are working in base two or *binary* base and so on. Each base uses the same number of digits as its value, e.g. base two uses two digits, base three uses three digits, base four uses four digits and so on. Each base uses all digits lower than its value, e.g. base two uses digits 0 and 1; base three uses digits 0, 1 and 2; base four uses digits 0, 1, 2 and 3.

3.1 Bases

Study the following table to understand some commonly used number bases, their names and digits they use.

Base	Name of base	No. of digits bases use	List of digits used
Two	Binary	2	0, 1
Five	Quinary	5	0, 1, 2, 3, 4
Eight	Octal	8	0, 1, 2, 3, 4, 5, 6, 7
Ten	Denary (Decimal)	10	0, 1, 2, 3, 4, 5, 6, 7, 8, 9
Twelve	Duodecimal	12	0, 1, 2, 3, 4, 5, 6, 7, 8, 9, t, e

Fig. 3.1

Bases can be represented as; 111 base two $\rightarrow$ 111_2 or 111_{two}

321 base eight $\rightarrow$ 321_8 or 321_{eight}

t027e base twelve $\rightarrow$ $t027e_{12}$ or $t027e_{twelve}$

Remember that *t* is a single digit for 10 and *e* is a single digit for 11.

Activity 3.1

Given bases, three, nine and eleven, draw a table to:
  (a) Give the number of digits each base has
  (b) List the digits in each case

3.2   Counting in bases

We can have numbers counted in bases, using a single digit or a group of digits. Following is a table showing the counting in various bases. Study the following table carefully.

Base	N U M B E R															
	1st	2nd	3rd	4th	5th	6th	7th	8th	9th	10th	11th	12th	13th	14th	15th	16th
Ten	0	1	2	3	4	5	6	7	8	9	10	11	12	13	14	15
Five	0	1	2	3	4	10	11	12	13	14	29	21	22	23	24	30
Eight	0	1	2	3	4	5	6	7	10	11	12	13	14	15	16	17
Twelve	0	1	2	3	4	5	6	7	8	9	t	e	10	11	12	13

Fig. 3.2

Do you realise the following equivalencies as we count?

$$9_{ten} \equiv 14_{five} \equiv 11_{eight} \equiv 9_{twelve}$$
$$12_{ten} \equiv 22_{five} \equiv 14_{eight} \equiv 10_{twelve}$$
$$15_{ten} \equiv 30_{five} \equiv 17_{eight} \equiv 13_{twelve}, \text{ etc}$$

Continue the above table up to the 20th number, then let us try to count in twos as we record. (Remember; we only have digits 0 and 1 to use in base 2. Copy and complete the next 16 numbers in base two in the counting below:

0, 1, 10, 11, 100, __ , __ , __ , __ , __ , __ , __ , __ , __ , __ , __ , __ , __ , __ , __ .

Activity 3.2

(a) Tabulate the first ten numbers for the   following bases;
  (i) base ten   (ii) base three   (iii) base six.
(b) From your table find the equivalencies of;   (i) 8_{ten} in base six
  (ii) 11_{three} in base six   (iii) 12_{three} in base six   (iv) 3_{six} in base three

---

3.3 Conversion of bases

Previously we have looked at various equivalences. Did you get the equivalence; $17_{ten} \equiv 32_{five} \equiv 21_{eight} \equiv 15_{twelve}$?

Considering the above equivalence we can convert 17_{ten} to base five to obtain $17_{ten} \equiv 32_{five}$. Similarly if you are given 32_{five} to convert it to base ten you will obtain 17_{ten}.

3.3.1 Conversion from base ten to other bases

Given 17_{ten} to convert it to base five we have; 17 sticks grouped in lots of fives:

///////////////// ⇨ /////   /////   /////   //
                        Lot 1   *Lot 2*   *Lot 3*   *Remainder (2)*

This gives 3 lots of 5s remainder 2 sticks, that is, 32. $\therefore$ **$17_{ten} = 32_{five}$.**

Similarly given 14_{ten} to convert it to base three we have; 14 sticks grouped in lots of threes:

/////////////// ⇨ ///   ///   ///   ///   //
                        Lot 1   *Lot 2*   *Lot 3*   *Lot 4*   *Remainder (2)*

This gives 4 lots of 3s remainder 2 sticks. 4 is greater than 3 (the base), so divide it further in 3s until you obtain the number of lots less than 3.

Let **/** represent ///, so we now have; 4 lots of 3s further divided by 3:

/ / / / ⇨ **/ / /**   **/**
               Lot 1   *Remainder (1)*

Giving 1 lot of 3, remainder 1 stick.

We finally have 1 lot, 1 stick and 2 sticks (as read in reverse direction starting from final number of lots obtained to remainders), that is, 112.

$\therefore$ **$14_{ten} = 112_{three}$**

Similarly, 9_{ten} to base two, we have;

///////// ⇨ //   //   //   //   /
                    Lot 1   *Lot 2*   *Lot 3*   *Lot 4*   *Remainder (1)*
                                               4 lots of 2s remainder 1

Let **/** represent //, so we now have; 4 lots of 2s further divided by 2:

| | Lot 1 | Lot 2 | Remainder (0) |

2 lots of 2s remainder 0

Let ▮ represent ▮ ▮, so we now have; 2 lots of 2s further divided by 2:

| Lot 1 | Remainder (0) |

1 lot of 2, remainder 0

We now have 1 lot, 0 sticks, 0 sticks and 1 stick as you read moving in the reverse direction. That is, 1001. $\therefore 9_{ten} = 1001_{two}$

Remember that we write the last lot once they are less than the base, followed by the remainders in reverse order for example the last remainder obtained becomes the first while writing the answer.

Activity 3.3

Convert the following bases as required using the above studied method (Make sure you understand the technique) :

1 12_{ten} to base four   2 15_{ten} to base six   3 11_{ten} to base two.

The previously learnt method of converting from base ten to other bases would be very tedious and unsuitable for larger numbers. In this case we have to resort to the following method;

Example 1

Convert the following as required: 321_{ten} to base four

Solution

4	3	2	1	R
4		8	0	1
4		2	0	0
4			5	0
			1	1

Keep diving the base ten number by 4 (the base to be converted to) and writing the remainder (**R**) in each case until you obtain a value less than the base.

Now read the answer in direction as shown by the arrow.

$\therefore 321_{ten} = 11001_{four}$

Example 2

Convert the following as required:

(i) 2702_{ten} to base eleven

(ii) 5062_{ten} to base twelve

Solution

(i)

11	2	7	0	2	R
11		2	4	5	7
11			2	2	3
				2	0

So, $2702_{ten} = 2037_{eleven}$

(ii)

12	5	0	6	2	R
12		4	2	1	t
12			3	5	1
				2	e

So, $5062_{ten} = 2e1t_{twelve}$

Activity 3.4

Convert the following bases as required using the approach used in the given examples:

(i) 17_{ten} to base five   (ii) 14_{ten} to base three   (iii) 9_{ten} to base two

Compare this method with the previous one in relation to what is happening as you divide.

Exercise 3a

1 10_{ten} to base two

2 23_{ten} to base three

3 182_{ten} to base four

4 113_{ten} to base five

5 217_{ten} to base six

6 412_{ten} to base seven

7 221_{ten} to base eight

8 1412_{ten} to base nine

9 2489_{ten} to base eleven

10 1463_{ten} to base twelve

3.3.2 Conversion of any base to base ten

Given 3104_{ten} we can represent it on an abacus with place values in powers of ten, that is, 10^0 (or 1), 10^1 (or 10), 10^2 (or 100), 10^3 (or 1000), etc.

Notice that on the abacus the place values are occupied by powers of ten.

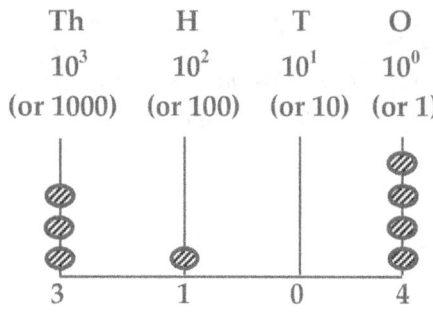

Expanded as:
$$(3 \times 10^3) + (1 \times 10^2) + (0 \times 10^1) + (4 \times 10^0)$$
$$3000 \quad + \quad 100 \quad + \quad 0 \quad + \quad 4$$

We can also represent 32_{five} on an abacus as:

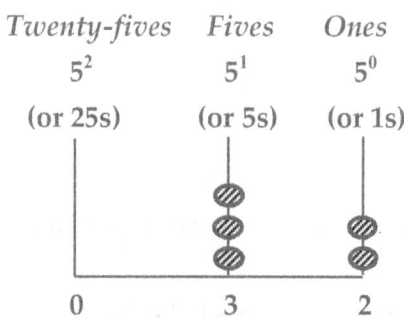

Expanded as: $(0 \times 5^2) + (3 \times 5^1) + (2 \times 5^0)$
$$0 \quad + \quad 15 \quad + \quad 2$$
$$= 17$$

This means that: $32_{five} = 17_{ten}$; and therefore we can use this approach to convert from any base to base ten.

Similarly given 2105_{six} to be converted to base ten we shall have;

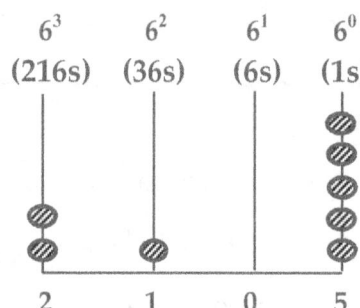

Expanded as:
$$(2 \times 6^3) + (1 \times 6^2) + (0 \times 6^1) + (5 \times 6^0)$$
$$(2 \times 216) + (1 \times 36) + (0 \times 6) + (5 \times 1)$$
$$432 \quad + \quad 36 \quad + \quad 0 \quad + \quad 5$$
$$= 473$$

Therefore, $2105_{six} = 473_{ten}$

Activity 3.5

Using an abacus convert the following bases to base ten: (i) 382_{nine}
(ii) 10096_{eleven}

We have now known how the place values in terms of the given base affect the digits of this base. We therefore do not necessarily need to use the abacus at this stage

Example 1

Convert the following bases to base ten: (i) 101402_{five} (ii) $10t1_{twelve}$

Solution

(i) Write down a series for the sum of each digit multiplied by the base (five) and give the place values by assigning the base (five) powers starting from zero as they increase towards the left.

(1×5^5) + (0×5^4) + (1×5^3) + (4×5^2) + (0×5^1) + (2×5^0)

(1×3125) +   0   + (1×125) + (4×25) +   0   + (2×1)

3125 +   0   +   125   +   100   +   0   +   2

$= 3352_{ten}$

Therefore, $101402_{five} = 3352_{ten}$

(ii) (1×12^3) + (0×12^2) + $(t \times 12^1)$ + (1×12^0)

1728 +   0   +   120   +   1

$= 1849_{ten}$

Therefore, $10t1_{twelve} = 1849_{ten}$

In a case where the number to be converted to base ten is a decimal fraction, you will always award powers following a number line. That is, from the given number the first digit after the decimal point takes the power $^{-1}$, the second takes $^{-2}$, the third takes $^{-3}$, and so on.

For example; $34.123_{five} \rightarrow (3 \times 5^1) + (4 \times 5^0) + (1 \times 5^{-1}) + (2 \times 5^{-2}) + (3 \times 5^{-3})$ or

$5.2703_{nine} \rightarrow (5 \times 9^0) + (2 \times 9^{-1}) + (7 \times 9^{-2}) + (0 \times 9^{-3}) + (3 \times 9^{-4})$, etc.

Example 2

Convert 1001.1101_{two} to base ten.

Solution

$(1 \times 2^3) + (0 \times 2^2) + (0 \times 2^1) + (1 \times 2^0) + (1 \times 2^{-1}) + (1 \times 2^{-2}) + (0 \times 2^{-3}) + (1 \times 2^{-4})$

8 +   0   +   0   +   1 $+ (1 \times \frac{1}{2}^1) + (1 \times \frac{1}{2}^2) + (0 \times \frac{1}{2}^3) + (1 \times \frac{1}{2}^4)$

8 + 0 + 0 + 1 + $\frac{1}{2}$ + $\frac{1}{4}$ + 0 + $\frac{1}{16}$

9 + $\frac{1}{2}$ + $\frac{1}{4}$ + 0 + $\frac{1}{16}$

$$9 + \frac{(8+4+1)}{16} \quad = \quad 9 + \frac{13}{16} \quad = 9\,{}^{13}/_{16}\text{ ten}$$

$$\therefore\ 1001.1101_{two} = 9\,{}^{13}/_{16}\text{ ten}\ (\textbf{or } 9.8125_{ten})$$

Exercise 3b

Convert the following bases as required

1 1106_{nine} to base ten

2 241_{five} to base ten

3 55_{eight} to base ten

4 24_{twelve} to base ten

5 540_{six} to base two

6 1176_{eight} to base twelve

7 1101_{two} to base five

8 1034_{five} to base eight

9 872_{nine} to base twelve

10 $10et_{twelve}$ to base eleven

3.4 Operations with bases

Example 1

Work out: $11101_{two} + 1110_{two}$

Solution

```
  1   1
  1 1 1 0 1 two
+   1 1 1 0 two
1 0 1 0 1 1 two
```

(1) $1 + 0 = 1$, write 1, since it is less than the base 2.

(2) $0 + 1 = 1$, write 1, since it is less than the base 2.

(3) $1 + 1 = 2$, we do not write 2 in base two, so divide the answer 2 by base 2 to obtain 1 r 0, write the remainder 0 and carry 1.

(4) 1 (*carried*) $+ 1 + 1 = 3$, we do not write 3 in base two, so we have $3 \div 2$ (base) $= 1$ r 1; write 1 the remainder and carry 1.

Example 2

Work out: $2614_{eight} + 4421_{eight}$

Solution

```
  1
  2 6 1 4 eight
+ 4 4 2 1 eight
  7 2 3 5 eight
```

(1) $4 + 1 = 5$, 5 is less than 8, so it is put directly as the answer. Like wise $1 + 2 = 3$, so it is directly put as the answer.

(2) $6 + 4 = 10$, $10 \div 8$ (base) $= 1$ r 2, write 2, carry 1.

Example 3

Work out: (i) $8176_{nine} - 1123_{nine}$ (ii) $4310_{five} - 2420_{five}$.

Solution

(i)

$$
\begin{array}{r}
8\ 1\ 7\ 6_{nine} \\
-\ 1\ 1\ 2\ 3_{nin} \\
\hline
7\ 0\ 5\ 3_{nine}
\end{array}
$$

(ii)

$$
\begin{array}{r}
{\scriptstyle 3\ \ 2} \\
4\ 3\ 1\ 0_{five} \\
-\ 2\ 4\ 2\ 0_{five} \\
\hline
1\ 3\ 4\ 0_{five}
\end{array}
$$

(1) $1 - 2$ is *"not possible"*, then borrow 1 in terms of the base (five) to have $(1 + 5) = 6$; then $6 - 2 = 4$.

(2) $2 - 4$, is *"not possible"*, then borrow 1 in terms of the base (five) to have $(2 + 5) = 7$, then $7 - 4 = 3$.

Activity 3.6

Work out: (i) $71t62_{eleven} + 3249_{eleven}$ (ii) $10011_{two} - 111_{two}$

Exercise 3c

Work out:

1   $230_{eight} + 1412_{eight}$

2   $1011_{two} + 11101_{two}$

3   $2004_{five} + 3310_{five}$

4   $2012_{three} + 1000_{three} + 1102_{three}$

5   $t0e1_{twelve} + 124t0_{twelve}$

6   $8213_{nine} - 5001_{nine}$

7   $3412_{six} - 2002_{six}$

8   $1249_{eleven} - 32t1_{eleven}$

9   $3613_{nine} + 134_{five}$ (answer in base ten)

10   $3112_{four} + 23_{six}$ (answer in base eight)

Example 4

Work out: (i) $324_{five} \times 11_{five}$ (ii) $416_{eight} \times 12_{eight}$

Solution

(i)

$$
\begin{array}{r}
3\ 2\ 4_{five} \\
\times\ \ \ 1\ 1_{five} \\
\hline
3\ 2\ 4_{five} \\
+\ 3\ 2\ 4\ 0_{five} \\
\hline
4\ 1\ 1\ 4_{five}
\end{array}
$$

(ii)

$$
\begin{array}{r}
4\ 1\ 6_{eight} \\
\times\ \ \ 1\ 2_{eight} \\
\hline
1\ 0\ 3\ 4_{eght} \\
+\ 4\ 1\ 6\ 0_{eight} \\
\hline
5\ 2\ 1\ 4_{eight}
\end{array}
$$

(1) $6 \times 2 = 12$, $\underline{12} = 1\ r\ 4$. write 4 carry 1.
    $\ \ \ \ \ \ \ \ \ \ \ 8$

(2) $1 \times 2 = 2$, $2 + 1$ (carried) $= 3$

(3) $4 \times 2 = 8$, $\underline{8} = 1\ r\ 0$.
    $\ \ \ \ \ \ \ \ \ \ 8$

Continue and add same in base, eight.

∴ $324_{five} \times 11_{five} = 4114_{five}$

∴ $416_{eight} \times 12_{eight} = 5214_{eight}$

Activity 3.7

(i) Convert 324$_{five}$ and 11$_{five}$ each to base ten then carryout the multiplication in base ten. Finally convert your answer to base five

(ii) Multiply 3021$_{four}$ by 23$_{four}$

Exercise 3d

Multiply:

1 321$_{four}$ x 12$_{four}$ 2 1011$_{two}$ x 100$_{two}$

3 1102$_{six}$ x 33$_{six}$ 4 5201$_{eight}$ x 23$_{eight}$

5 1234$_{five}$ x 20$_{five}$ 6 11011$_{two}$ x 111$_{two}$

7 1001$_{two}$ x 23$_{four}$ (answer in base ten)

8 213$_{six}$ x 12$_{eight}$ (answer in base twelve)

REVISION EXERCISE 3

1 Convert the following bases from base ten to other bases as required;
  (i) 102$_{ten}$ to base two   (ii) 247$_{ten}$ to base six   (iii) 1941$_{ten}$ to base twelve

2 Convert the following bases as given to base ten.
  (i) 101011$_{two}$ (ii) 10034$_{eight}$ (iii) 208t1$_{eleven}$ (iv) 101.011$_{two}$

3 Arrange the following numbers in ascending order
  (Hint: first convert the given group of numbers into the same base);
  (i) 323$_{four}$, 110111$_{two}$, 43$_{ten}$, 44$_{five}$ (ii) 56$_{twelve}$, 509$_{ten}$, tt$_{eleven}$, 66$_{eight}$.
  (iii) 15$_{twelve}$, t$_{eleven}$, 1111$_{two}$, 40$_{six}$ (iv) 42$_{eleven}$, 30$_{twelve}$, 11$_{eleven}$, 505$_{six}$

Work out the following

4 11$_{two}$ + 101$_{two}$ 5 1011$_{two}$ + 11111$_{two}$

6 1001$_{two}$ + 11$_{two}$ 7 314$_{five}$ + 104$_{five}$

8 372$_{nine}$ + 22$_{nine}$ 9 227$_{eleven}$ + 344$_{eleven}$

10 111$_{two}$ − 10$_{two}$ 11 336$_{eight}$ − 321$_{eight}$

12 1398$_{twelve}$ − 420$_{twelve}$ 13 431$_{five}$ − 32$_{five}$

14      $3106_{eleven} - 1215_{eleven}$ 15 $20t_{twelve} - 11e_{twelve}$

Multiply as required.

16      $3021_{five} \times 211_{five}$ 17 $1251_{six} \times 23_{six}$

18      $326_{eight} \times 21_{eight}$ 19 $424_{twelve} \times 42_{twelve}$

20   Work out the following:

   (i) $1011_{two} - 101_{two} + 111_{two}$

   (ii) $307_{eight} \times 5_{eight} + 162_{eight}$

22 Workout the following and give your answer as indicated in the bracket to the right of the question.

   (i) $53_{six} \times 32_{four} + 113_{four}$ (base ten)

   (ii) $121_{eight} - 32_{eight} \times 11_{ten}$ (base eight)

22 Write each of the following expressions as expansions in base n:

   (i) $2n^2 + 2n + 4$ (ii) $3n^3 + 2n + 4$

23 Find the value of n in each of the following
   (i) $23_n = 201_{three}$ (ii) $302_n = 62_{nine}$

24 Each of the following equations are in base eight, find the value of x in base eight in each case.
   (i) $x - 16 = 7$ (ii) $12 - x = 6$

25   Given that $31_p = 26_{eight}$, find the value of p.

4 NUMERICAL CONCEPTS

INTRODUCTION

Numerical concepts constitute further number ideas. It covers how numbers relate under various operations. Under a number of instances, you will be revisiting the use of the four basic operations of arithmetic.

4.1 Multiples, Factors and Divisors

4.1.1 Factors

If we have numbers multiplied together, say 2 x 3 we get an answer 6. 2 and 3 are factors of 6.

Similarly; 2 x 3 x 4 = 24 meaning that 2, 3 and 4 are factors of 24.

When a natural number is divided by its factor it gives an answer which is exact (whole), also a factor of this same number. To find the factors of a number say 24, you take all the numbers that exactly go into 24 (or those which leave no remainder when they divide 24). Let us try to check for the factors of 24;

1   is a factor of 24, since it goes into 24 exactly twenty-four times.

2   is a factor of 24, since it goes into 24 exactly twelve times.

3   is a factor of 24, since it goes into 24 exactly eight times.

4   is a factor of 24, since it goes into 24 exactly six times.

5   is Not a factor of 24, since it leaves a remainder, 4 when it divides 24.

6   is a factor of 24, since it goes into 24 exactly four times.

7   is Not a factor of 24, since it leaves a remainder, 3 when it divides 24.

8   is a factor of 24, since it goes into 24 exactly three times.

9   is Not a factor of 24, since it leaves a remainder, 6 when it divides 24.

10   is Not a factor of 24, since it leaves a remainder, 4 when it divides 24.

11   is Not a factor of 24, since it leaves a remainder, 2 when it divides 24.

12   is a factor of 24, since it goes into 24 exactly two times.

24 is a factor of 24, since it goes into 24 exactly once.

$$\therefore \text{ The factors of 24 are; } \quad F_{24} = \{1, 2, 3, 4, 6, 8, 12, 24\}$$

Always remember that:

(1)   1 is a factor of every number

(2)   Every number is a factor of itself

(3)   Proper factors are those factors of a number excluding the number itself. For example, the proper factors of 24 are {1, 2, 3, 4, 6, 8, 12}.

Example 1

Find the factors of 42, 72, 108 and 120 by checking the numbers which go into them exactly.

Solution

F_{42} = {1, 2, 3, 6, 7, 14, 21, 42}

F_{72} = {1, 2, 3, 4, 6, 8, 9, 12, 18, 24, 36, 72}

F_{108} = {1, 2, 3, 4, 6, 9, 12, 18, 27, 36, 54, 108}

F_{120} = {1, 2, 3, 4, 5, 6, 8, 10, 12, 15, 20, 24, 30, 40, 60, 120}

Activity 4.1

Find and list the factors of: (i) 18   (ii) 32   (iii) 45   (iv) 64
Comment on the factors of 32 in comparison with those of 64.

Common Factors

When we look through the factors of 42, 72, 108 and 120 in example 1 above, we find out that some factors are common in these different curl brackets. For example the common factors for 42, 72, 108 and 120 are {1, 2, 3, 6}.
If F_{12} ={1, 2, 3, 4, 6, 12}, F_{18} ={1, 2, 3, 6, 9, 18} and F_{90} = {1, 2, 3, 5, 6, 9, 15, 18, 90}.

- The common factors of 18 and 90 are $F_{18} \cap F_{90}$ = {1, 2, 3, 6, 9, 18}.

- The common factors of 12, 18 and 90 are $F_{12} \cap F_{18} \cap F_{90}$ = {1, 2, 3, 6}.

Greatest/Highest Common Factors (GCF/HCF)

From the list of factors for 42, 72, 108 and 120 in example 1 above, we have common factors as; {1, 2, 3, 6}. We can then tell that the Greatest/Highest Common Factor is 6.

Method for finding GCFs

We can follow a particular method to find the GCF of any group of numbers.

Example 2

Find the GCF of 72, 90 and 108.

Solution

2	72	90	108
3	36	45	54
3	12	15	18
	4	5	6

Use the least prime factor which goes into all the numbers. Divide until no more primes can divide all the numbers.

Then multiply together all the prime factors used to divide the numbers

$\therefore$ GCF of 72, 90 and 108 $\quad = \quad 2 \times 3 \times 3$

$= \quad 18$

Example 3

Find the GCF of   (i) 30, 42, 72 and 120   (ii) 80, 300 and 540

Solution

(i)

2	30	42	72	120
3	15	21	36	60
	5	7	12	20

(ii)

2	80	300	540
2	40	150	270
5	20	75	135
	4	15	27

$\therefore$ GCF of 30, 42, 72 and 120 $= 2 \times 3$

$= 6$

$\therefore$GCF of 80, 300 and 540 $= 2 \times 2 \times 5$

$= 20$

Activity 4.2

By listing all the factors of 80, 300 and 540, find their GCF. Compare your answer with that in part (ii), example 3 above.

Exercise 4a

Find the factors of the following numbers.

1	388	2	43	3	180	4	250
5	1080	6	480	7	555	8	2134

9 1111   10 7073

Find the common factors of the following sets of numbers by listing factors.

11 80, 160 and 20   12 34, 64 and 200

13 2, 4, 6 and 20   14 63, 77 and 350

15 20, 102 and 480

Find the GCFs for the following sets of numbers using the discussed method.

16 36, 90 and 108   17 25, 120 and 150

18 44, 50, 90 and 1000   19 205, 335, 115 and 1050.

20 460, 23, 253 and 529.

4.1.2 Multiples

We have earlier on looked at factors multiplied to give a certain natural number. This natural number is known as a multiple. For example, 2 x 3 forms a multiple 6, of both 2 and 3.

We can list multiples of 5 as;

$$(5 \times 1), (5 \times 2), (5 \times 3), (5 \times 4), (5 \times 5), (5 \times 6), (5 \times 7) \ldots$$

That is;   5,   10,   15,   20,   25,   30,   35 …

M_5 = {5, 10, 15, 20, 25, 30, 35 …}, where M_5 represents the multiples of 5.

We can also have multiples of 2, 3 and 4 as: M_2 = {2, 4, 6, 8, 10, 12, 14 …}

M_3 = {3, 6, 9, 12, 15, 18, 21…}

M_4 = {4, 8, 12, 16, 20, 24, 28…}

Common multiples

When we study carefully the multiples of 2, 3 and 4; we shall be able to list the common multiples of 2, 3 and 4. These are {12, 24, 36, 48,…}

Similarly, if M_6 = {6, 12, 18, 24, 30, 36 …} and M_9 = {9, 18, 27, 36, 45, 54 …}

Then the common multiples of 6 and 9 are $M_6 \cap M_9$ = {18, 36 … }.

Lowest/Least Common Multiples (LCMs)

From the lists of multiples of 2, 3 and 4 we find out that the common multiples are {12, 24, 36, 48...}. Therefore, the Lowest/Least Common Multiple of 2, 3 and 4 is 12.

Finding the Lowest/Least Common Multiples (LCMs)

It is very tiresome to find the LCM by listing the multiples for every number and trace for the lowest especially for large numbers. We therefore resort to the use of the following method.

Example 1

Find the LCM of 40, 88 and 100

Solution

2	40	88	100
2	20	44	50
2	10	22	25
5	5	11	25
5	1	11	5
11	1	11	1
	1	1	1

Divide throughout by prime factors starting with the smallest which can exactly go into any of the three numbers. Carry on the number which is not divisible by the prime factor (11 in this case), until a stage when it is divided by the next prime used for which it is divisible.

∴ The LCM of 40, 88 and 100 is: 2 x 2 x 2 x 5 x 5 x 11

= 2200

Example 2

Find the LCM of: (i) 32, 40, 72 and 80

                     (ii) 60, 88, 90, 102, and 120.

Solution

Follow the same procedure as in example 1 above, and of course not forgetting to carry on any of the numbers which is not divisible by the dividing prime factor.

Work out as follows next page.

(i) 32, 40, 72 and 80

2	32	40	72	80
2	16	20	36	40
2	8	10	18	20
2	4	5	9	10
2	2	5	9	5
3	1	5	9	5
3	1	5	3	5
5	1	5	1	5
	1	1	1	1

∴ the LCM of 32, 40, 72 and 80 is;

2 x 2 x 2 x 2 x 2 x 3 x 3 x 5

= 1440

(ii) 60, 88, 90, 102, and 120.

2	60	88	90	102	120
2	30	44	45	51	60
2	15	22	45	51	30
3	15	11	45	51	15
3	5	11	15	17	5
5	5	11	5	17	5
11	1	11	1	17	1
17	1	1	1	17	1
	1	1	1	1	1

∴ The LCM of 60, 88, 90, 102 and 20 is;

2 x 2 x 2 x 3 x 3 x 5 x 11 x 17

= 67320

Exercise 4b

By listing common multiples, find the LCMs of the following sets of numbers

1 6, 9 and 15   2 5, 8, 12 and 15   3 3, 5 and 7

Using the discussed method, find the LCMs of the following sets of numbers

4 40, 22 and 8   5 11, 15, 21, and 30   6 30, 100 and 48

7 5, 8, 7 and 9   8 95, 15, 30 and 45

9 100, 125, 300 and 120   10 17, 30, 270 and 510

4.1.3 Divisors

When we divide 2 numbers, say 12 ÷ 3 = 4; in this mathematical statement, 3 is called the divisor. 12 is the dividend and 4 is the quotient. Remember that in this case 3 and 4 are factors of 12, and 12 is a multiple of both 4 and 3.

Divisibility Tests

Before dividing numbers, we can first check whether the divisor goes exactly into the dividend. This helps us to easily workout the division. This process of checking is known as the *divisibility test*.

Study carefully the following divisibility tests:

- A number is divisible by 2 if it is even (or if its last digit is even).
- A number is divisible by 3 if the sum of its digits is a multiple of three.
- A number is divisible by 4 if its last two digits form a multiple of 4.
- A number is divisible by 5 if its last digit is 0 or 5.
- A number is divisible by 6 if it is even and the sum of its digits is a multiple of 3.
- A number is divisible by 8 if the number formed by the last 3 digits of that number is a multiple of 8.
- A number is divisible by 9 if the sum of its digits is a multiple of 9.
- A number is divisible by 10 if its last digit is zero.
- A number is divisible by 11 if the difference between the sum of the digits in even places and the sum of digits in the odd places is zero or is a multiple of 11.
- A number is divisible by 12 if it is divisible by both 3 and 4.

Exercise 4c

State whether the numbers below are divisible by; 3, 4, 5, 6, 9, 10, 11 and 12.

1 1248	2 40453	3 4800	4 210348
5 4444	6 101013	7 8118	8 2180
9 4245	10 5556		

4.1.4 Applications of GCFs and LCMs

We can use GCFs and LCMs to solve problems in daily life.

Example 1

Three bells ring after every 30 minutes, 40 minutes and 50 minutes respectively. If all the bells ring at the same time after how many minutes will they ring again at the same time?

Solution

1st bell rings after every 30 minutes

2nd bell rings after every 40 minutes

3rd bell rings after every 50 minutes

The time at which they will all ring at the same time will be obtained by finding the LCM of 30, 40 and 50.

The LCM of 30, 40 and 50 is; $2 \times 2 \times 2 \times 3 \times 5 \times 5 = 600$

$\therefore$ the three bells will ring again at the same time after 600 minutes.

Example 2

Lines are drawn so as to divide a rectangle 50cm long by 30cm wide into equal squares. Find the least possible number of squares.

Solution

This can be obtained from the GCF of 50cm and 30cm

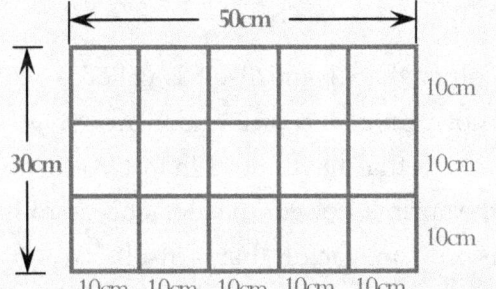

$F_{50} = \{1, 2, 5, 10, 25, 50\}$

$F_{30} = \{1, 2, 3, 5, 6, 10, 15, 30\}$

Use **10**, the GCF to divide sides-length and width.

We then obtain 5 by 3 = 15 squares of 10cm side each.

Exercise 4d

1 In a factory a clay brick is produced after every 125 seconds and a cement brick is produced after every 75 seconds. If a clay brick and a cement brick are produced at the same time, after how many minutes will they again be produced at the same time?

2 A patient takes tablets of type A after every 3 hours, tablets of type B after every 4 hours and tablets of type C of tablets at the same time.

(a) After how many hours will he again take all the three types of tablets at the same time?

(b) After how many hours will he take all the three types of tables at once for the third time?

3 Find the least sum of money which can be divided into an equal number of shares either of Sh.60 or Sh.80 or Sh.200 each.

4 A rectangular block 18cm by 36cm by 45cm is cut into an exact number of cubes. Find the least possible number of cubes.

5 Either by striding 75cm or 80cm I take an exact number of steps to walk across the court-yard. Find the least width of the court-yard.

6 Find the least sum of money such that the remainder is Sh. 5 when it is divided by 10, 15 or 18.

4.2 Prime and Composite Numbers

A whole number that has no factors other than 1 and itself is called a *prime number*. All whole numbers that are not prime, except 0 and 1 are *composite numbers*. Composite numbers are numbers that have more than two distinct factors. The numbers 0 and 1 are neither prime nor composite, since zero has an endless number of factors and 1 has only one factor, that is, itself.

Given the sets, we have;   Whole, $W = \{0, 1, 2, 3, 4, 5, 6, 7, 8, ...\}$,

Prime, $P = \{2, 3, 5, 7, 11, ...\}$ and

Composite, $C = \{4, 6, 8, 9, 10, 12, ...\}$

Have you realized that each of the prime numbers in set P has no factors except 1 and itself? Whereas, the composite numbers have more than 2 distinct factors, say 6 has factors 1, 2, 3 and 6. Try to find the factors of other composite numbers in the set C.

4.2.1 Finding Prime numbers

Let us consider finding all the prime numbers between 1 and 50.

Step 1: Write down the natural numbers to be searched without missing out any as shown below.

Step 2: Cross out 1, (use " X "). 1 is not prime.

Step 3: Leave 2, but circle it (2 is prime) and cross every multiple of 2, that is, 4, 6, 8, etc. (use " / ")

Step 4: Leave 3, but circle it (3 is prime) and cross every multiple of 3, that is, 9, 15, 21, etc. which is not crossed yet. (use " — ")

Step 5: 4 is already crossed out, so leave 5, but circle it (5 is prime) and cross out every multiple of 5, that is not crossed— 25 and 35. (use " | ")

Step 6: 6 is already crossed out, so leave 7, but circle it (7 is prime) and cross out every multiple of 7, that is not crossed, that is, 49. (use " \ ")

Step 7: 8, 9 and 10 are already crossed out, so leave 11, but circle it (11 is prime) and then trace for multiples of 11 that are not crossed. In this case all the multiples of 11 are already crossed. We can then stop here and take all the uncrossed numbers to be prime.

X	②	③	4̸	⑤	6̸	⑦	8̸	9̶	10̸
⑪	1̸2	13	14̸	1̶5̶	1̸6	17	1̸8	19	20̸
2̶1̶	2̸2	23	2̸4	2̸5	2̸6	2̶7̶	2̸8	29	3̸0
31	3̸2	3̶3̶	3̸4	3̸5	3̸6	37	3̸8	3̶9̶	4̸0
41	4̸2	43	4̸4	4̶5̶	4̸6	47	4̸8	4̸9	5̸0

This is an algorithm for finding prime numbers based on the *Eratosthenes' sieve*. Eratosthenes was a Greek mathematician.

Therefore the prime numbers between 1 and 50 are;

$$\{2, 3, 5, 7, 11, 13, 17, 19, 23, 29, 31, 37, 41, 43, 47\}$$

Exercise 4e

1 Use Eratosthenes' sieve to find the prime numbers between 1 and 200.

2 Find the prime numbers between (i) 58 and 90   (ii) 10 and 41

(iii) 22 and 41   (iv) 31 and 37.

3 Find the sum of the prime numbers between (i) 28 and 39   (ii) 19 and 23.

4 State *true* or *false*.

(i) All prime numbers are odd.

(ii) All composite numbers are even

(iii) 1 is neither prime nor composite

(iv) There are 4 prime numbers less than 10.

Example 1

Find the prime numbers which are factors of; (i) 68   (ii) 144.

Solution

First find the factors of the given numbers.

(i) F_{68} = {1, 2 4, 17, 34, 68}   ∴ the prime factors of 68 are; {2, 17}

(ii) F_{144} = {1, 2 , 3 , 4, 6, 8, 9, 12, 16, 18, 24, 36, 48, 72, 144}

∴ The prime factors of 144 are; {2, 3}.

Example 2

Find three consecutive prime numbers which add up to: (i) 41   (ii) 131.

Solution

Obtain the average of the numbers;

(i) $\dfrac{41}{3}$ = 13 r 2   13 is prime, so let's check the next prime numbers on its either sides.

Above 13 ⇒ 17

At average ⇒ 13

Below 13 ⇒ 11   **Check:** Is 11 + 13 + 17 = 41? <u>Yes</u>.

∴ 11, 13 and 17 are the three consecutive prime numbers that add up to 41.

(ii) $\dfrac{131}{3}$ = 43 r 5   43 is prime, so we have;

$$41 \xleftarrow{} 43 \xrightarrow{} 47$$

Check: Is 41 + 43 + 47 = 131? <u>Yes</u>

∴ 41, 43 and 47 are the three consecutive prime numbers that add up to 131.

Example 2

Find four consecutive prime numbers that add up to 72.

Solution

Find the average; $\dfrac{72}{4}$ = 18.   If the average is not prime, we take the nearest or immediate prime number to it.

It could be below or above the average. For 18 as the average, we have the next immediate prime number: **(i)**   Below 18 is 17 and

(ii)   Above 18 is 19.

Now we have;

$$13 \xleftarrow{} 17, 19 \xrightarrow{} 23$$

Check: Is 13 + 17 + 19 + 23 = 72? <u>Yes</u>

∴ the 4 consecutive prime numbers that add up to 72 are; 13, 17, 19 and 23.

Exercise 4f

1 Find two consecutive prime numbers whose sum is 84.

2 Find the prime numbers which are factors of
   (i) 88   (ii) 72   (iii) 120   (iv) 210.

3 Find three consecutive prime numbers which add up to   (i) 289 (ii) 83

4 Find four consecutive prime numbers that add up to;   (i) 120   (ii) 378

5 Find five consecutive prime numbers that add up to;   (i)   67

                                                            (ii)   161

6 Find the sum of composite number between   (i)   20 and 50

                                                 (ii)   33 and 41

4.2.1 Prime factorizing

A composite number can be expressed as a product of prime factors e.g. 10 can be written as 2 x 5; 36 can be written as 2 x 2 x 3 x 3 etc. Prime factorizing of a composite number involves dividing it by prime numbers, 2, 3, 5, 7, 11, 13 … in this order using a factor tree.

Example 1

Prime factorize 1080

Solution

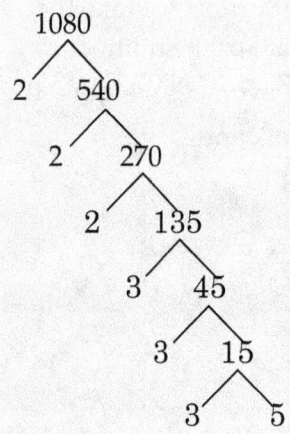

Divide 1080 by the lowest prime number which leaves a remainder zero after the division.

Until the value to be divided is prime then stop, e.g. in this case we stop at five 5, a prime number.

$$1080 = 2 \times 2 \times 2 \times 3 \times 3 \times 3 \times 5$$

$$\therefore 1080 = 2^3 \times 3^3 \times 5$$

Example 2

Prime factorize:　　(i) 729　　　(ii) 2448

Solution

(i) 729　　　　　　　(ii)

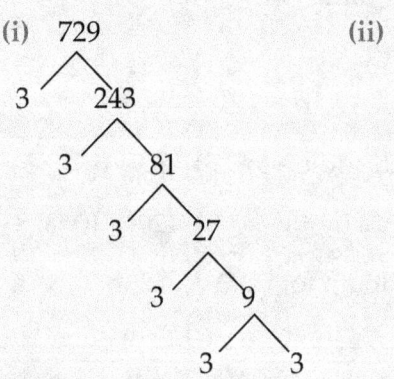

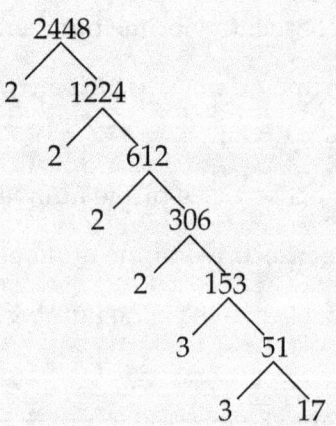

$729 = 3 \times 3 \times 3 \times 3 \times 3 \times 3$　$\therefore 729 = 3^6$　　　　$2448 = 2 \times 2 \times 2 \times 2 \times 3 \times 3 \times 17$

$$\therefore 2448 = 2^4 \times 3^2 \times 17$$

Prime factorize each of the following   numbers.

1 11048	2 4809	3 1102904	4 10005
5 111432	6 40048	7 330334	8 9196

Listing prime factors

We can represent prime factors of a number as a set. When we prime factorize 1080, we obtain 1080 = 2 x 2 x 2 x 3 x 3 x 3 x 5. These are *three* 2s, *three* 3s and *one* 5.

We can then have a list of prime factors of 1080 as: $\{2_1, 2_2, 2_3, 3_1, 3_2, 3_3, 5_1\}$.

Example 2

List the prime factors of 2100

Solution

First prime factorize to obtain: 2100 = 2 x 2 x 3 x 5 x 5 x 7

∴ the prime factors of 2100 are; $\{2_1, 2_2, 3_1, 5_1, 5_2, 7_1\}$

Example 3

List the prime factors of 10164

Solution

First prime factorize to obtain: 10164 = 2 x 2 x 3 x 7 x 11 x 11

∴ the prime factors of 2100 are; $\{2_1, 2_2, 3_1, 7_1, 11_1, 11_2\}$

Exercise 4g

Prime factorize each of the following numbers and list its prime factors.

1 1250	2 2115	3 1039
4 2940	5 15750	6 29575

4.2.2   Using Venn diagrams to find GCFs

We can represent lists of prime factors of different numbers as sets on Venn diagrams. We can then use the Venn diagram interpretations to find the GCF of the number whose prime factors are represented.

Example 1

Find the GCF of 126, 315 and 420 using Venn diagrams.

Solution

First prime factorize to obtain:

$126 = 2 \times 3 \times 3 \times 7 \qquad \Rightarrow \{2_1, 3_1, 3_2, 7_1\}$

$315 = 3 \times 3 \times 5 \times 7 \qquad \Rightarrow \{3_1, 3_2, 5_1, 7_1\}$

$420 = 2 \times 2 \times 3 \times 5 \times 7 \Rightarrow \{2_1, 2_2, 3_1, 5_1, 7_1\}$

On a Venn diagram;

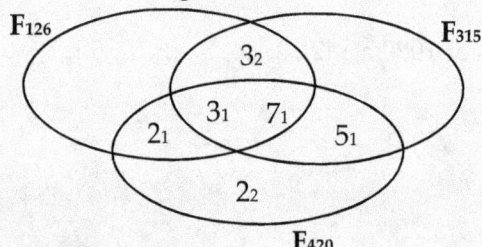

$F_{126} \cap F_{315} \cap F_{420} = \{3_1, 7_1\}$

∴ the GCF of 126, 315 and 420 is:

$3 \times 7 = 21$

Activity 4.3

From example 5 above, the GCF of 126 and 315 is $F_{126} \cap F_{315}$ giving $\{3_1, 3_2, 7_1\}$ and leading to $3 \times 3 \times 7$; which is equal to 63.

Using the same Venn diagram find the GCF of   (i) 126 and 420

                                                  (ii) 315 and 420

Exercise 4h

Using Venn diagrams, find the GCFs of;

1 420, 600 and 800   2 920, 1200 and 1500

3 54, 90 and 300   4 72, 150 and 240

5   640, 720 and 2480

4.3   Even and Odd numbers

Whole numbers can be categorized as either even or odd. The even and odd numbers keep alternating continuously as shown on the number line (*fig. 4.1, next page*). Please carefully study the relationship between the odd and even numbers on the illustration that follows.

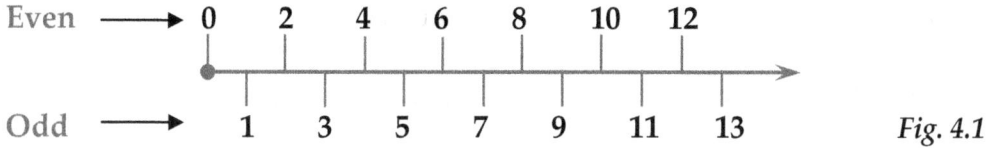

Fig. 4.1

4.3.1 Even numbers

Given that we have a set of whole numbers, W, such that W = {0, 1, 2, 3 ...}, if we multiply each whole number in set W by 2 up to n whole numbers, we shall obtain a set of results, E as follows;

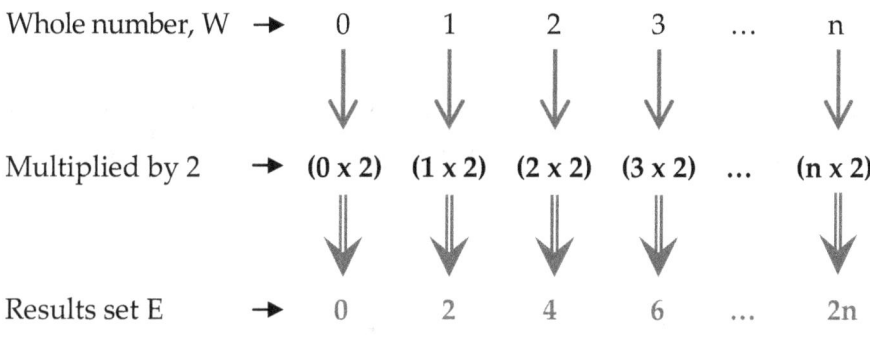

Fig. 4.2

The results in set E are numbers each twice its corresponding whole number in set W. A number that is 2 times a whole number is an *even number*. Therefore E is a set of even numbers that is, E = {0, 2, 4, 6, 8, 10 ... 2n}. An even number can generally be represented as $2n$ where n is any whole number. An even number is also known as *"a number divisible by 2 and leaves no remainder"*.

4.3.2 Odd numbers

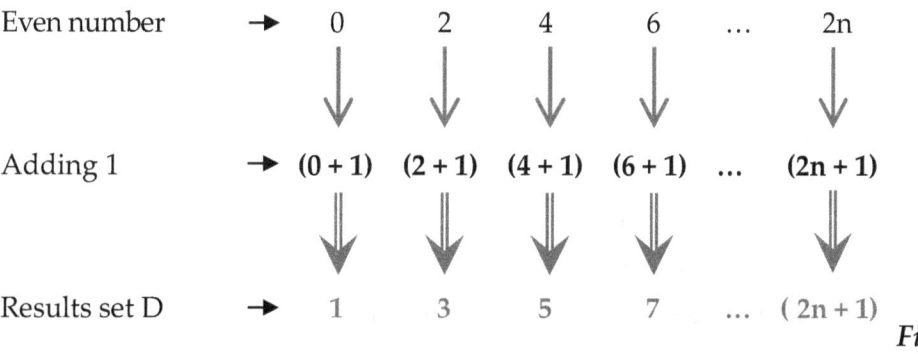

Fig. 4.3

From *fig. 4.3* on the previous page, if we add 1 to each of the numbers in set E (set of even numbers), we shall have a results set D.

∴ set D = {1, 3, 5, 7, 9, 11 ... (2*n* + 1)}. Each of the numbers in set D is an even number plus 1. set D is a set of odd numbers. Therefore *odd numbers are numbers that exceed even numbers by 1.* Odd numbers can be represented generally as (2*n* + 1), where *n* is any whole number and 2*n* is an even number. An odd number is also known to be *"a number which when divided by 2 leaves 1 as a remainder".*

4.3.3 Operations with Prime, Even and Odd numbers

When dealing with operations on prime, even and odd numbers, you may predict the nature of your result before actually working it out. This would help you to make a pre-check to guide you for your answerꞨ nature at a glance.

Addition

Odd + Odd	= Even:	For example; 21 + 13 = 34
Even + Even	= Even:	For example; 10 + 6 = 16
Even + Odd	= Odd:	For example; 8 + 5 = 13

Multiplication

Odd x Odd	= Odd:	For example; 7 x 3 = 21
Even x Even	= Even:	For example; 4 x 6 = 24
Even x Odd	= Even:	For example; 8 x 11 = 88

Remember that all prime numbers are odd except 2.

Activity 4.3

The sum of 2 prime numbers is 50;
(i) Find the possible prime numbers that we can add to get 50.
(ii) Is the product of these prime numbers even or odd? Why?

Exercise 4*i*

1   List all the odd numbers between 0 and 100

2   List all the even numbers between 0 and 100.

3   List all the even prime numbers

4   List all the odd prime numbers between 0 and 20

5   The sum of four consecutive odd numbers is 124. find the numbers.

6   The sum of three consecutive odd numbers is 99. Find the numbers.

4.4   Integers

In chapter 2, we looked at uses of numbers and one of them was to show some things in a reference system. Temperature is a measure in degrees which requires a reference frame that begins with a zero point and has an interval for scale (Fahrenheit or Celsius). Calibrations of these temperature scales is an application of integers.

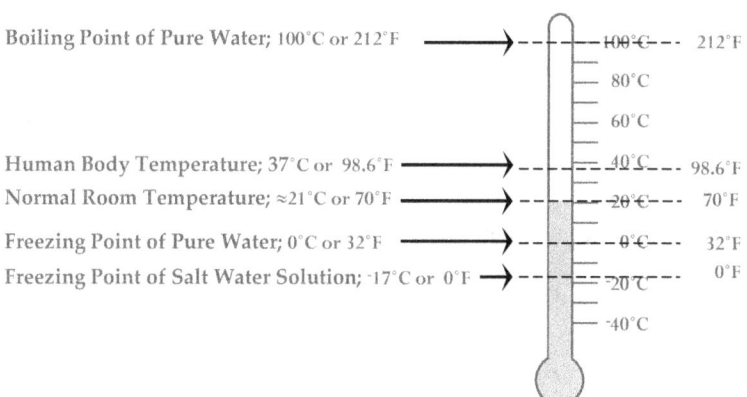

Thermometer

Fig. 4.4

Integers include positive numbers, negative numbers and the number zero. They are also known to be *directed* or *signed* numbers.

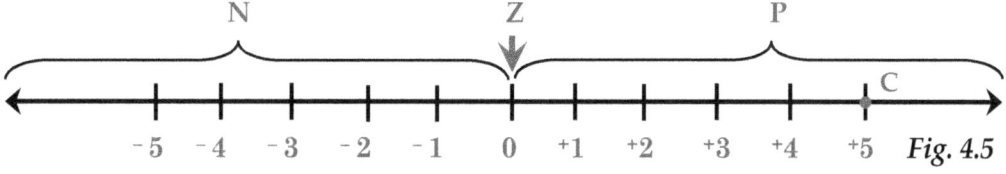

Fig. 4.5

Integers can be better illustrated as shown on the number line on the previous page; *fig.4.5.*

A number line starts and continues with equally spaced points from zero to its either sides.

N is a negative region and it is on the left of zero.

P is a positive region and it is on the right of zero.

Z is a point called the origin and is marked Zero. The counting begins here.

The coordinate of point C is $^+5$, so we can have coordinates $^+1$, $^+2$, $^+3$ ...; as positive integers and $^-1$, $^-2$, $^-3$ ...; as negative integers.

4.4.1 Properties of integers

On the number line the value of integers increases as we move towards the right (positive region) and decreases towards the left (negative region). For example, ... $^-2 < ^-1 < 0 < ^+1 < ^+2$ ...

Have you realized that for every positive integer there is also a negative integer with the same distance from the origin? For example, $^-2$ and $^+2$, $^-5$ and $^+5$, $^-9$ and $^+9$ etc. A pair of such integers as $^-1$ and $^+1$ is called *opposites*. Similarly, each of the pairs $^-2$ and $^+2$, $^-5$ and $^+5$, $^-17$ and $^+17$, etc. are opposites and 0 is its own opposite.

Another name for opposite is *additive inverse*. Therefore, the additive inverse of $^-1$ is $^+1$, etc.

From the number line we realize two vital properties required to describe integers, that is

(i)   the direction from zero (the origin)   and

(ii)   the distance from zero (the origin)

The direction is represented by a sign, that is, "positive, $(^+)$", for the direction towards the right of the origin, and "negative, $(^-)$", for the direction towards the left of the origin. Remember that zero is neither positive nor negative; and a positive number may be represented with or without a sign e.g. $^+8$ can be represented as 8.

The distance of any integer from zero is called the *absolute value (magnitude)* of that integer, say if R stands for any real number, then the absolute value of R is $|R|$.

For example, $|{}^+4| = 4$ and $|{}^-4| = 4$. Therefore, $|{}^+4| = |{}^-4| = 4$.

Have you realized that from the number line $^+4$ is 4 units from zero towards the right hand side of the number line? And that $^-4$ is 4 units from zero to the left hand side of the number line?

Remember: Do **NOT** confuse *the value of an integer* and *the absolute value of an integer*. From the number line in *fig. 4.5*, page 75; we have:

$^+5 > {}^+4$, $^+2 > 0$, $0 > {}^-1$, and so on.

Activity 4.4

Fill in either =, > or < :

1 0 ___ ⁻10   2 3 ___ ⁻3   3 ⁻7 ___ ⁻5   4 ⁻6 ___ ⁻6

Exercise 4j

Draw a number line with 10 units on either sides of Zero and use it to answer the following questions.

1   Write down the first 4 negative even numbers.

2   Write down the first 4 negative odd numbers

3   Write down the first 4 negative prime numbers.

4   Which number is neither positive nor negative?

5   Find the opposites of   (i) ⁻4   (ii) ⁺3   (iii) 5
                                     (iv) 0   (v) 12

6   Find the absolute values of   (i) ⁻8   (ii) ⁻1   (iii) ⁺5
                                     (iv) 0   (v) 15

For the numbers 7 and 8 do not use the number line.

7 (a) List the positive even numbers from 0 to 20.

   (b) List the negative prime numbers from 0 to ⁻20.

8   Find the absolute values for the following

   (i)   11   (ii) ⁺23   (iii)   ⁻43   (iv) ⁻89   (v) 69

4.4.3 Addition of integers

By direction signs model

If opposites are added you obtain zero as the answer. For example, $^+13 + (^-13)$ gives 0, $^+7 + (^-7) = 0$, $^+1 + (^-1) = 0$ etc. If $^+1 + (^-1) = 0$, then each negative unit cancels out with a positive unit on addition. Study the illustrations that follow;

Breaking down integers to units, we have:

$^+5 = (^+1) + (^+1) + (^+1) + (^+1) + (^+1)$ $^+3 = (^+1) + (^+1) +(^+1)$

$^-5 = (^-1) + (^-1) + (^-1) + (^-1) + (^-1)$ $^-4 = (^-1) + (^-1) + (^-1) + (^-1)$

Diagrammatically we have,

(i) $^+5 + ^-5 = 0$

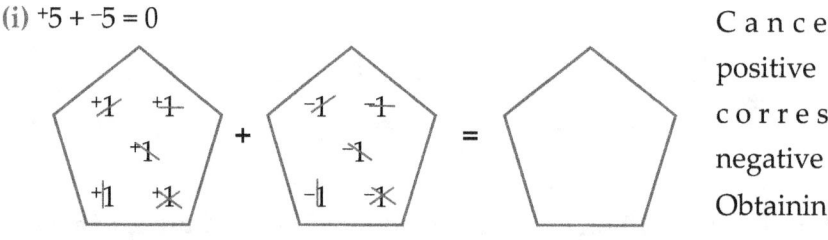

Cancel each positive unit with a corresponding negative unit;

Obtaining: $^+5 + ^-5 = 0$

(ii) $^+5 + ^-4 = ^+1$ (iii) $^-5 + (^+3) = ^-2$

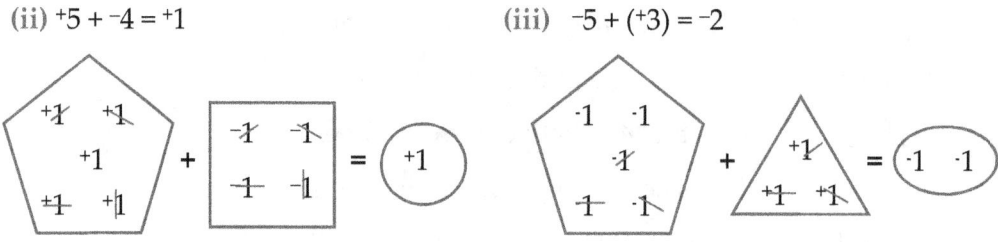

This method is only suitable for small value integers and basic practice. However, for further understanding of harder problems proceed to other methods of integer addition which follow.

Activity 4.5

Using the above discussed method (*Direction signs model*) illustrate the following problems

1 $^-5 + (^+3) + (^-4) + (^+6)$ 2 $^+3 + (^-3) + (^-4) + (^+5)$

3 $^-7 + (^-4) + (^+6)$ 4 $^-4 + (^+5) + (^-3) + (^+7)$

By number line model

Example 1

Add: ⁺5 + (⁻8)

Solution

Using zero as the origin move 5 units to the right (⁺5), then 8 units (⁻8) to the left from the value at which you are, (i.e from ⁺5). Now read the value at which you have ended as your answer, that is, ⁺5 + (⁻8) = ⁻3

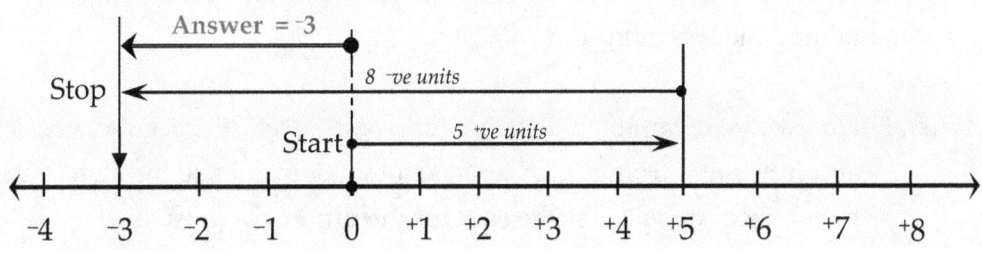

Example 2

Add: ⁺2 +(⁺3) + (⁻4)

Solution

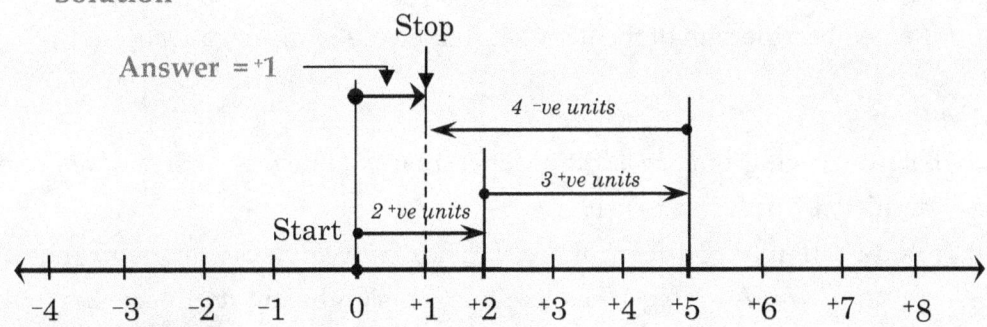

Activity 4.6

1 Add (i) ⁺5 + (⁻5) =   (ii) ⁻3 + (⁺3) =   (iii) ⁺8 + (⁻8) =

2 On a number line show that: (i)   ⁻5 + (⁺3) + (⁻4) + (⁺6) = 0

                                      (ii) ⁻9 + (⁺5) + (⁻3) + (⁺7) = 0

By calculations

To add integers we shall put into consideration the distance and direction from zero. We shall use 2 rules in order to carry out the addition operation on integers

Rule I: For two real numbers with the *same sign*,

[that is, if both are positives (⁺) or if both are negatives (⁻)];

we add their absolute values and use their common sign as the sign of the sum (answer).

For example;

(i) From (⁺3) + (⁺5), we have: $|{}^+3| + |{}^+5| = (3 + 5) = 8$. So, (⁺3) + (⁺5) = ⁺8. Maintaining the common sign.

(ii) From (⁻2) + (⁻3), we have: $|{}^-2| + |{}^-3| = (2 + 3) = 5$. So, ⁻2 + (⁻3) = ⁻5 Maintaining the common sign.

Rule II: For two real numbers with *opposite signs*, we subtract the smaller absolute value from the larger absolute value and use the sign of the larger absolute value as the sign for the difference (answer).

For example;

(i) From (⁻2) + (⁺3), we have: $|{}^+3| - |{}^-2| = (3 - 2) = 1$. So (⁻2) + (⁺3) = ⁺1 Remember, the sign of the larger absolute value , 3 is positive(⁺).

(ii) From (⁺4) + (⁻7), we have: $|{}^-7| - |{}^+4| = (7 - 4) = 3$. So, (⁺4) + (⁻7) = ⁻3 Remember, the sign of the larger absolute value , 7 is negative(⁻).

Remember:-

1 Before carrying out an addition operation on integers, first analyze and decide on which rule to use.

2 Study carefully and learn the difference between a *direction sign* and an *operation sign*. An integer has its direction sign before itself, e.g ⁺2, ⁻3, ⁺4, etc. Remember a number with out a sign represents a positive number.

The illustration below explains the direction and operation signs.

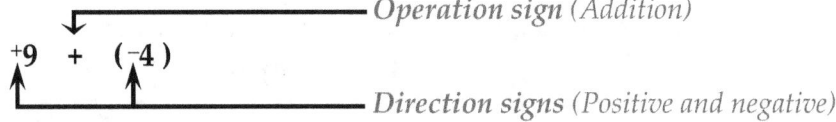

Operation sign *(Addition)*

⁺9 + (⁻4)

Direction signs *(Positive and negative)*

Activity 4.7

Underline all the direction signs and circle all the operation signs.

1 ⁺4 + −11 − ⁺5 + −3 + −6 − −7   2 −13 + −11 − −9 + ⁺6 + 7 − 8

Example 3

Add: (i) ⁺4 + (⁺9)   (ii) ⁻2 + (⁻5)

   Solution

(i) Here we shall use the first rule for addition of integers (for like signs);

$$| ^+4 | \rightarrow 4$$
$$| ^+9 | \rightarrow \underline{9}$$
     Sum   <u>13</u>   The common sign is positive, so use a positive
                         sign for sum; Therefore,   ⁺4 + (⁺9) = **⁺13**

(ii) Here we shall use the first rule for addition of integers (for like signs);

$$| ^-2 | \rightarrow 2$$
$$| ^-5 | \rightarrow \underline{5}$$
     Sum   <u>7</u>   The common sign is negative, so use a negative
                         sign for a sum obtained;
                         Therefore,   ⁻2 + (⁻5) = **⁻7**

Example 2

Add: (i) ⁻7 + (⁺11)   (ii) ⁺6 + (⁻10)

   Solution

(i) Here we use the second rule for addition of integers (for different signs);

$$| ^-7 | \rightarrow 7$$
$$| ^+11 | \rightarrow \underline{11}$$
   Difference   <u>4</u>   For addition of different sign integers we get
                       the difference and it takes the sign of the larger
absolute value. The difference takes the positive sign. Therefore, ⁻7 + (⁺11) = **⁺4**

(ii) Here we use the second rule for addition of integers (for different signs);

$$| ^+6 | \rightarrow 6$$
$$| ^-10 | \rightarrow \underline{10}$$
   Difference   <u>4</u>   For different signs, we get the difference of
                       absolute values and it takes the sign of the
                       larger absolute value. Therefore, ⁺6 + (⁻10) = **⁻4**

Exercise 4k

Using the idea of *"opposite units canceling,"* in numbers 1 to 5, work out:

1 ⁻6 + (⁺9)   2   ⁻7 + (⁺5) + (⁺2)

3 ⁻6 + (⁻8) + (⁺1)   4   ⁻2 + (⁺1) + (⁻4) + (⁻2) + (⁺3)

5 ⁻4 + (⁺5) + (⁻4) + 9 + (⁺3) + (⁻1)

In numbers 6 to 20 perform the following additions and use the number line illustration to check your answers.

6 $^-2 + (^+3)$

7 $^+5 + (^-5)$

8 $7 + (^-6)$

9 $^-9 + (^+10)$

10 $^+6 + (^+7) + (^+10)$

11 $^-9 + (^-6) + (^+7)$

12 $^+8 + (^-6) + (^+5)$

13 $^-11 + (^+5) + (^-2)$

14 $^-7 + (^+6) + (^-4)$

15 $^-9 + (^+7) + (^+9)$

16 $^-1 + (^-5) + (^+6)$

17 $^-8 + (^+7) + (^+5)$

18 $^+11 + (^-1) + (^-12)$

19 $^-7 + (^+6) + (^+8) + (^-11)$

20 $^-6 + (^+5) + (^+7) + (^-6)$

4.4.4   Subtraction of integers

We now know how to add integers. We are therefore going to use the earlier learnt addition method to subtract integers.

Since we can define subtraction in terms of addition, we shall learn one rule in subtraction and the rest will be the addition rules as given and practiced earlier.

Rule: For all real numbers a and b, $a - b = a + (^-b)$.

This means that *"subtracting b from a"* is the same thing as *"adding a to the opposite of b"*, that is, $a + (^-b)$", which we have already looked at earlier in addition of integers.

For example;
(i) $^+5 - (^+3) = {}^+5 + (^-3) = {}^+2$ *"Minus ($^+3$) = Plus ($^-3$)"*

(ii) $^+4 - (^+1) = {}^+4 + (^-1) = {}^+3$ *"Minus ($^+1$) = Plus ($^-1$)"*

You will realize that in order to subtract an integer from another, you simply add the opposite of the integer being subtracted to the first one.

(i) $^+4 - (^+2) = {}^+4 + (^-2)$ *minus $^+2$*

         plus opposite of $^+2$ } *minus $^+2$ = plus opposite of $^+2$*

         $= {}^+2$

(ii) $^+5 - (^-2) = {}^+5 + (^+2)$ *minus $^-2$*

         plus opposite of $^-2$ } *minus $^-2$ = plus opposite of $^-2$*

         $= {}^+7$

Have you realized how the signs change? By now it is not necessary to write a positive number with a sign, for example, $^+5$, $^+7$, $^+8$, etc. are simply written as 5, 7, 8, etc. respectively.

Example 1

Subtract: (i) $2 - 5$ (ii) $^-8 - 5$ (iii) $^-3 - (^-4)$

Solution

(i) $2 - 5 = 2 + (^-5)$ (ii) $^-8 - 5 = ^-8 + (^-5)$ (iii) $^-3 - (^-4) = ^-3 + (^+4)$

$= ^-3$ $= ^-13$ $= 1$

Example 2

Work out: (i) $^-2 - 5 + 8 - 1$ (ii) $3 - (^-1) + 3 - 5$

Solution

Combine all the negatives and positives separately to get the answer as follows:

(i) $^-2 - 5 + 8 - 1$

$= ^-2 + (^-5) + 8 + (^-1)$

$= (^-2) + (^-5) + (^-1) + 8$

$= (^-8) + 8$

$= 0$

(ii) $3 - (^-1) + 3 - 5$

$= 3 + 3 - 5 - (^-1)$

$= 3 + 3 + (^-5) + (^+1)$

$= 7 + (^-5)$

$= 2$

Exercise 4I

Subtract the following:

1 $5 - 4$ 2 $3 - (^-6)$ 3 $^-2 - 4$

4 $^-11 - (^-2)$ 5 $^-5 - 10 - 3$ 6 $^-1 - (^-6)$

7 $7 - (^-9)$ 8 $^-2 - 9 - (^-10)$ 9 $^-22 - 10$

10 $^-5 - (^-5)$ 11 $23 - (^-1) - 30$ 12 $^-17 - 8 - (^-10)$

If you work out more problems in addition and subtraction, you will gain a skill of performing subtraction problems without addition of the opposite sign in order to get the answer. You can then perform the addition, subtraction or a combination of them without using the previous steps but rather getting the answers directly.

4.4.5 Multiplication of integers

Obtaining products for integers is done in the same way as for the whole numbers, however, here we shall only consider how the direction signs change.

In multiplication of integers we shall use the following rules:

Rule I: If the two factors have *like signs*, then the sign of the product is *positive* that is;

$$(^+a) \times (^+b) = ^+(ab) \quad \text{For example;} \quad (^+3) \times (^+4) = ^+12$$

$$(^-a) \times (^-b) = ^+(ab) \quad \text{For example;} \quad (^-3) \times (^-4) = ^+12, \text{ etc.}$$

Rule II: If the two factors have *different signs*, then the sign of the product is *negative*, that is;

$$(^+a) \times (^-b) = ^-(ab) \quad \text{For example;} \quad (^+3) \times (^-4) = ^-12$$

$$(^-a) \times (^+b) = ^-(ab) \quad \text{For example;} \quad (^-3) \times (^+4) = ^-12, \text{ etc.}$$

These two rules will help us to workout all the problems involving multiplication of integers.

Example 1

Multiply: (i) $(^+3) \times (^-2) \times (^-6)$ (ii) $(2) \times (^-4) \times (10) \times (^-1)$

Solution

Here note that we handle two at ago.

(ii) $(2) \times (^-4) \times (10) \times (^-1)$

(i) $(^+3) \times (^-2) \times (^-6)$

$(^-8) \times (10) \times (^-1)$

$(^-6) \times (^-6)$

$(^-80) \times (^-1)$

$= 36$

$= 80$

Exercise 4m

Multiply in the following numbers

1 $2 \times (^-3)$ 2 $^-2 \times (^-9)$ 3 3×8

4 $^-7 \times 6$ 5 $^-9 \times (^-10)$ 6 $5 \times (^-11)$

7 $^-11 \times (^-1)$ 8 $0 \times (^-11)$ 9 $^-4 \times 9 \times (^-5)$

10 $12 \times (^-10) \times (^-2)$

4.4.6   Division of integers

To divide integers, we only need to consider how direction signs change.

Rule I: The quotient of two integers with *like signs* is *positive*, that is;

$$(^+a) \div (^+b) = ^+(^a/_b)$$ For example; $(^+14) \div (^+2) = ^+7$

$$(^-a) \div (^-b) = ^+(^a/_b)$$ For example; $(^-14) \div (^-2) = ^+7$

Rule II: The quotient of two integers with *unlike signs* is *negative*, that is;

$$(^+a) \div (^-b) = ^-(^a/_b)$$ For example; $(^+14) \div (^-2) = ^-7$

$$(^-a) \div (^+b) = ^-(^a/_b)$$ For example; $(^-14) \div (^+2) = ^-7$

Example 1

Divide: (i) $8 \div 2$ (ii) $^-10 \div 5$ (iii) $15 \div (^-3)$

Solution

(i) $8 \div 2 = 4$ (ii) $^-10 \div 5 = ^-2$ (iii) $15 \div (^-3) = ^-5$

Exercise 4n

Divide the following

1 $18 \div 2$	2 $^-20 \div 4$	3 $^-5 \div (^-5)$
4 $100 \div (^-20)$	5 $^-60 \div (^-5)$	6 $75 \div (^-15)$
7 $[55 \div (^-11)] \div ^-5$	8 $[^-72 \div (^-8)] \div ^-3$	
9 $9 \div [^-48 \div (^-8)]$	10 $[^-90 \div 45] \div ^-2$	

4.4.7   Combined Operations on Integers

More than one operation may be used on integers in a single statement, for example, given: $^-2(8 + 2)$, we can work it out as;

$$^-2(8 + 2) \quad = ^-2 (10) \qquad \text{Or} \qquad ^-2(8 + 2) \quad = ^-16 + ^-4$$
$$= ^-20 \qquad\qquad\qquad\qquad\qquad\quad = ^-20$$

And for ; $^-3 + [\, 5 - (^-1)\,]$, we resort to BODMAS, that is, $^-3 + [\, 5 - (^-1)\,]$,

$$^-3 + (5 + 1)$$
$$^-3 + 6$$
$$= 3$$

BODMAS is an acronym to help you remember the order of operation in case of mixed operations.

Example 1

Workout: (i) $^-2 + 2 [2 - (^-7)]$ (ii) $8 - 4 [(^-1) + 6]$

Solution

(i) $^-2 + 2 [2 - (^-7)]$ (ii) $8 - 4 [(^-1) + 6]$

$= ^-2 + 2 [2 + 7]$ $= 8 - 4 (5)$

$= ^-2 + 2 (9)$ $= 8 - 20$

$= ^-2 + 18$ $= ^-12$

$= 16$

Example 2

Work out: $1 + [(^-8) \div 2]$ of $(^-5) \times 4 - 5$

Solution

Start with **Brackets** → **Of** → **Multiplication** → **Addition** → **Subtraction**; as guided by underlines.

$1 + [(^-8) \div 2]$ of $(^-5) \times 4 - 5$ $= 1 + (\underline{^-8 \div 2})$ of $(^-5) \times 4 - 5$ *(Brackets)*

$= 1 + (\underline{^-4}) \text{ of } (^-5) \times 4 - 5$ *(Of)*

$= 1 + \underline{20 \times 4} - 5$ *(Multiplication)*

$= \underline{1 + 80} - 5$ *(Addition)*

$= \underline{81 - 5}$ *(Subtraction)*

$= 76$

Therefore, $1 + [(^-8) \div 2]$ of $(^-5) \times 4 - 5 = 76$

Exercise 4p

Work out the following

1   $3 + [2 - (^-6)]$ 2 $^-6 + 3[(^-3) + 2]$

3   $^-11 - 4[10 - (^-2)]$ 4 $80 - 11(2 - 3)$

5 8 x 4 – 2[10 ÷ (⁻2)]

6 ⁻6 + (⁻4) x 5 + [(⁻18) ÷ 3] of ⁻2

7 2 – 6[4 + (⁻7)] + 5 x 9

8 ⁻3 + 3(33 ÷ 11) + [3 – (⁻1)] of 4

9 [24 ÷ (⁻6)] x [4 – (⁻6)] – 6 – (⁻8)

10 $\dfrac{⁻6 – (⁻2) \times ⁻4}{(28 ÷ 7)}$

REVISION EXERCISE 4

1 Use the working of *Npposite units canceling'* to work out the following;

(i) ⁺7 + (⁻3)   (ii) ⁻1 + (⁻6) – (⁻7) – 3   (iii) 2(⁻3) + (⁺7) – 1

2 Illustrate how to obtain the solutions of the following integer operations on a number line.

(i) ⁺3 + (⁻4)   (ii) ⁻7 + (⁻2) – (⁺3)   (iii) 2(⁻3) + (⁺7) – 1

3 Cross check your answers for (i) and (ii) in 1 above using a number line.

4 Work out the following: (i) (⁻4) x 3 + (⁻5) x 4 ÷ (⁻2) (ii) $\dfrac{(⁻8 – ⁻4) \times ⁻3}{27 ÷ (⁻9)}$

5 (a)   Find the GCF of 42, 72, 108, and 240.

(b)   Find the LCM of 45, 25, 30 and 105.

(c)   Arrange digits 4, 2, 3, 8 and 5 to obtain a number (s) divisible by both 4 and 11.

6 (a)   The sum of 4 consecutive even numbers is 420. Find the numbers.

(b)   Find all the even numbers of which are multiples of 7 and are less than 100.

7 A bell rings after every 20 minutes, another one rings after every 45 minutes and the third one rings after every 60 minutes. If the bells ring at the same time now, after how many minutes will the three bells ring at the same time again?

8 In playing a certain game, if a player throws two dice and the sum of the 2 numbers shown up is odd and prime he gains 1 point; prime and even he gains 3 points; even and not prime he gains no point.

Write down all the possible combinations which would earn the player.

(i) 3 points

(ii) 1 point

(iii) No point

(Hint: use the table on the right)

+	1	2	3	4	5	6
1	2	3	4			
2						
3			6			
4						
5		7				
6						12

Complete this table and get combinations say (2, 5), which give the sum 7. 7 is an odd and prime number therefore a combination of 2 and 5 earns 1 point.

5 COMMON FRACTIONS

INTRODUCTION

Fractional numbers exist generally in the form a/b where a is called the *numerator* and b is called the *denominator*. For example; in a common fraction $2/5$, 2 is the numerator and 5 is the denominator. Fractions may be used to represent various situations in academics or real life;

5.1   Uses of Fractions

5.1.1   Representing one or more of the equal parts of a whole

If a unit is divided into 10 equal parts, then each part can be represented as a fraction of the whole or base unit. (A representative of one of the several equal parts of the base unit).

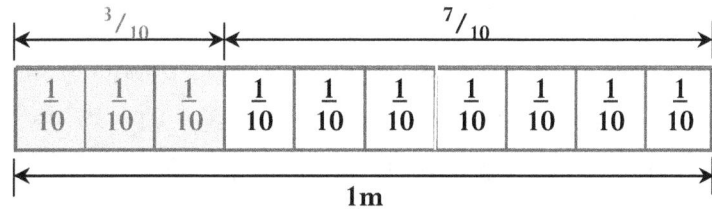

Fig. 5.1

From fig. 5.1 above, the base unit is 1m. It is divided into 10 equal parts. Each part is $1/10$ of a meter. 1 part represents $1/10$ of 1 meter.

Similarly;   2 parts represent $2/10$ of a meter ($2/10$ of the base unit).

3 parts represent $3/10$ of a meter ($3/10$ of the base unit) and so on.

3 out of 10 equal parts can be written as a fraction $3/10$, that is;

Numerator – considered parts (shaded)

$$\frac{3}{10}$$

Denominator – number of equal parts in a base unit as a whole (both shaded and unshaded parts added together)

Activity 5.1

1 Write the representative of one of the several equal parts of the base units, in each figure below;

(a) 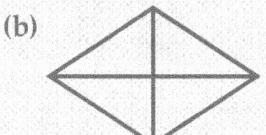   (b)   (c)

2 Write the fractions representing the shaded parts in the figures that follow.

(a) 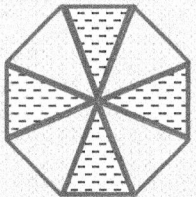   (b)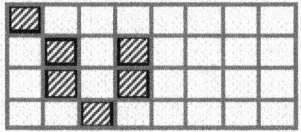

3 A bar of soap has six equal sections. If I use 2 sections from the bar, what fraction of the soap will I have used?

5.1.2 As representative of division

Consider 15 students sharing 5 cakes equally. For 15 students to share 5 cakes equally, the cakes must also be divided into 15 equal parts, therefore each cake must be divided into 3 equal parts, (that is, $^{15}/_5 = 3$)

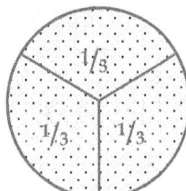 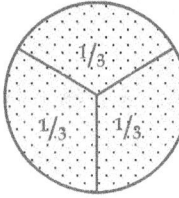 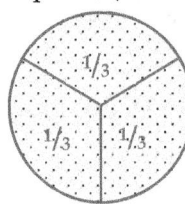 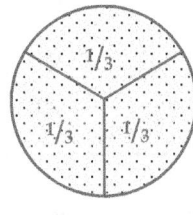 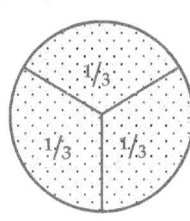

From above we have 15 students sharing 5 cakes equally each getting $^1/_3$ of a cake.

Activity 5.2

Shade the following divisions on diagrams showing how many several equal units each base unit has:

(i) $^{20}/_5$ (ii) $^{16}/_4$ (iii) $^9/_3$ (iv) $^{24}/_7$

5.1.3 Naming of points on a number line

Fractions can be used to name points between whole numbers on a number line.

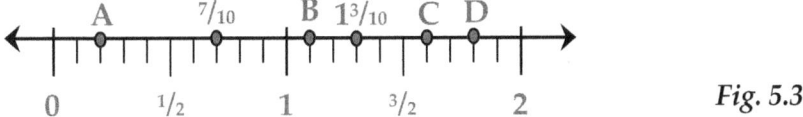

Fig. 5.3

Activity 5.3

Write the fraction naming for the points A, B, C and D as indicated on the number line in fig. 5.3 above.

5.1.4 As Ratios and Rates

When comparing two numbers with the same units, we use ratios. 12 is 3 times as large as 4. This means: 12 = 3 x 4. It still means that there are; three 4s (or four 3s) in 12. Therefore the ratio of 4 : 12 will be simplified in fractional form as;

$$\frac{4}{12} = \frac{1}{3} \qquad \text{or} \qquad \frac{3}{12} = \frac{1}{4}$$

(i.e. 4 : 12 :: 1 : 3) (i.e. 3 : 12 :: 1 : 4)

Rates are used to compare quantities with different units. A car can make 72 km on 14 litres of fuel. At the same rate, it can travel 36km on 7 litres.

$$\frac{72 \text{ km}}{14\,l} = \frac{36 \text{ km}}{7\,l}$$ These are equivalent fractions.

Activity 5.5

(a) Express the following fractions as ratios. (i) ½ (ii) ³/₉ (iii) ¹¹/₂₂

(b) Express the following ratios as fractions. (i) 1 : 3 (ii) 18 : 36 (iii) 10 : 15

5.1.5 Scales

Scales are used to compare the size of the actual object with its drawing or image. A map scale of 1 : 125 000 can be expressed as a fraction $^{1}/_{125\,000}$. This means that each unit on a map, represents 125 000 units on ground.

5.1.6 Probabilities

Fractions are a way of representing the chance that an event will occur. The probability that your village football team wins a match is $^1/_3$. This means that the chance of winning a match is only one out of the three possibilities; (that is, a win, draw or loss).

5.2   Categories of fractions

We can categorize a fraction depending on the nature of the numerator and denominator it has.

5.2.1   Proper and Improper fractions

Like whole numbers, fractions can too be represented on a number line. Considering 2 units on a number line, each divided into 6 equal parts, we have;

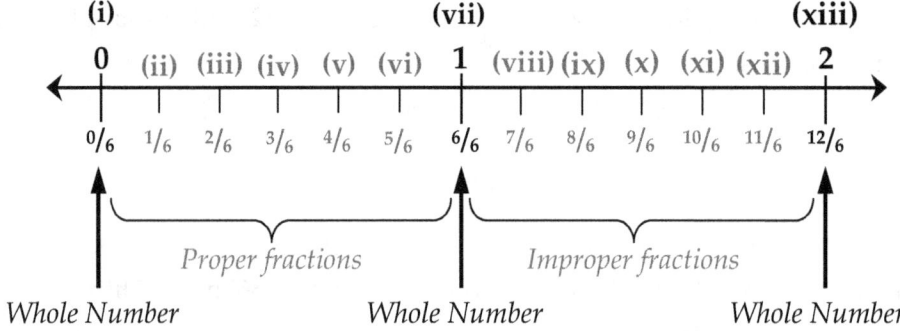

Fig. 5.4

Fractions which form whole numbers are those with the numerator equal to zero or those with the numerator which is a multiple of the denominator.

From the above number line we have;

$^0/_6 = 0$ (whole number); $^6/_6 = 1$ (whole number); $^{12}/_6 = 2$ (whole number), etc

Proper fractions

Proper fractions are those fractions with their numerators less than their denominators. For example, $^1/_6$, $^2/_6$, $^3/_6$, $^4/_6$, $^5/_6$, as shown in fig. 5.4.

Other examples of proper fractions include: $^3/_7$, $^1/_4$, $^2/_5$, $^9/_{11}$, etc.

1 What common characteristic (s) does the proper fractions have?

2 Think about and write down any ten proper fractions.

Improper fractions

Improper fractions are fractions with the numerator greater than the denominator. For example, the fractions $^7/_6$, $^8/_6$, $^9/_6$, $^{10}/_6$, $^{11}/_6$ as shown in the fig.5.4. Other examples of improper fractions include: $^8/_5$, $^{10}/_7$, $^4/_3$, $^3/_2$ etc.

5.2.2 Mixed numbers

A mixed number constitutes a whole number part and a proper fraction part. For example, $5\frac{1}{3}$, $2\frac{2}{3}$, $1\frac{4}{7}$, $3\frac{8}{11}$, etc. In this case we have say *"six whole numbers and a third"* for $6\frac{1}{3}$.

Have you realized from fig.5.4 that the point (xv) is 2 units plus $^2/_6$ of a unit? This point can be written as; $^{14}/_6$ (as an improper fraction) or as $2\frac{2}{6}$ (as a mixed fraction). Improper fractions can be expressed as mixed fractions and vice versa.

Example 1

Express as improper fractions: (i) $4\frac{2}{3}$ (ii) $2\frac{1}{8}$ (iii) $5\frac{2}{11}$

Solution

Here we multiply the denominator by the whole number; add the numerator to the result, then the final result is put over the denominator.

(i) $4\frac{2}{3} = 4 \to 2 = \dfrac{(3 \times 4) + 2}{3} = \dfrac{12 + 2}{3} = \dfrac{14}{3}$

(ii) $2\frac{1}{8} = 2 \to 1 = \dfrac{(8 \times 2) + 1}{8} = \dfrac{16 + 1}{8} = \dfrac{17}{8}$

(iii) $5\frac{2}{11} = 5 \to 2 = \dfrac{(11 \times 5) + 2}{11} = \dfrac{55 + 2}{11} = \dfrac{57}{11}$

Example 2

Express as improper fractions   (i) $\dfrac{14}{3}$ (ii) $\dfrac{10}{3}$ (iii) $\dfrac{13}{5}$

Solution

Remember fractions as representatives of division so in this case we divide the numerator by the denominator to get the answer as a whole number and the remainder will be expressed as a fraction of the denominator.

(i) $^{14}/_3$ = 4 whole numbers remainder 2 therefore; $^{14}/_3$ = $4\,^2/_3$

(ii) $^{10}/_3$ = 2 whole numbers remainder 1 therefore; $^{10}/_3$ = $2\,^1/_3$.

(iii) $^{13}/_5$ = 2 whole numbers remainder 3 therefore; $^{13}/_5$ = $2\,^3/_5$

Activity 5.7

1 Express the following mixed fractions as improper fractions.
  (i) $1\,^2/_3$ (ii) $2\,^1/_7$ (iii) $5\,^3/_4$ (iv) $7\,^1/_3$
2 Express the following improper fractions as mixed fractions.
  (i) $^{11}/_5$ (ii) $^{13}/_7$ (iii) $^{15}/_8$ (iv) $^{27}/_6$

4.2.3 Equivalent fractions

On your geometrical set ruler, separate marks in centimeters (cms) are represented. The centimeters (cms) are further divided into 10 equal units each called a millimeter (mm).

Carefully study the following illustration showing the comparison of the centimeter and millimeter units.

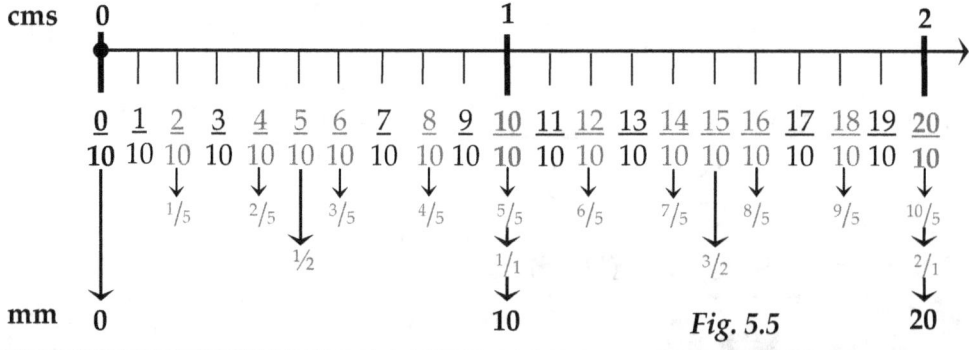

Fig. 5.5

From the illustration in fig.5.5, various fractions can represent the same value.

For example; $^8/_{10} \equiv {}^4/_5$, $\qquad$ $^{15}/_{10} \equiv {}^3/_2$, $\qquad$ $^{20}/_{10}, \equiv {}^{10}/_5$, $\qquad$ $^2/_1 \equiv 2$, and so on.

Have you realized that; $\qquad \dfrac{20}{10} = \dfrac{2 \times 10}{1 \times 10}$ $\qquad$ and $\qquad \dfrac{15}{10} = \dfrac{3 \times 5}{2 \times 5}$

Example 1

Write equivalent fractions for $^2/_7$

Solution

$$\frac{2}{7} = \frac{2 \times 2}{7 \times 2} = \frac{2 \times 3}{7 \times 3} = \frac{2 \times 4}{7 \times 4} \; ... \qquad \text{Thus;} \qquad \frac{2}{7} = \frac{4}{14} = \frac{6}{21} = \frac{8}{28} \; ...$$

Therefore the equivalent fractions to $\dfrac{2}{7}$ are; $\dfrac{4}{14}$, $\dfrac{6}{21}$, $\dfrac{8}{28}$ $\quad ...$

Example 2

Write equivalent fractions for $^3/_{11}$

Solution

$$\frac{3}{11} = \frac{3 \times 2}{11 \times 2} = \frac{3 \times 3}{11 \times 3} = \frac{3 \times 4}{11 \times 4} \; ... \quad \text{Thus;} \quad \frac{6}{22} = \frac{9}{33} = \frac{12}{44} = \frac{15}{55} \; ...$$

Therefore the equivalent fractions to $\dfrac{3}{11}$ are; $\dfrac{6}{22}$, $\dfrac{9}{33}$, $\dfrac{12}{44}$, $\dfrac{15}{55}$ $\quad ...$

Reducing fractions to lowest terms

Fractions can be expressed in a reduced form whereby the denominator and numerator have a common factor or when one is a multiple of the other, for example, $^{10}/_{20}$, $^3/_{18}$, $^8/_{20}$, $^5/_5$, etc. can be reduced to their lowest terms as ½, $^1/_6$, $^2/_5$, 1 etc. respectively.

Have you realized that; $\qquad$ (i) $\quad \dfrac{10}{20} = \dfrac{1}{2}$ (canceling by 10 to get $^1/_2$)

(ii) $\quad \dfrac{3}{18} = \dfrac{1}{6}$ (canceling by 3 to obtain $^1/_6$)

(iii) $\quad \dfrac{8}{20} = \dfrac{2}{5}$ (canceling by 4 to obtain $^2/_5$)

(iv) $\quad \dfrac{5}{5} = \dfrac{1}{1} = 1$ (canceling by 5 to obtain 1)

We can use the following methods to reduce fractions to lowest terms.

Reducing fractions to lowest terms — By prime factorizing

In $^{24}/_{36}$ we can factorize both the numerator and the denominator separately, that is,

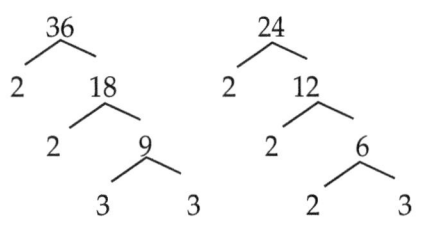

We now have; $\dfrac{24}{36} = \dfrac{2 \times 2 \times 2 \times 3}{2 \times 2 \times 3 \times 3}$

By canceling the common terms;

$$\dfrac{24}{36} = \dfrac{2 \times 2 \times 2 \times 3}{2 \times 2 \times 3 \times 3} = \dfrac{2}{3}$$

$36 = 2 \times 2 \times 3 \times 3$ $24 = 2 \times 2 \times 2 \times 3$ $\therefore\ ^{24}/_{36} = ^2/_3$ (in lowest terms).

Reducing fractions to lowest terms — By use of the GCF

For $^{30}/_{42}$ first find the GCF of 30 and 42

$F_{30} = \{ 1, 2, 3, 5, 6, 10, 15, 30\}$ and $F_{42} = \{1, 2, 3, 6, 7, 14, 21, 42\}$.

From the factors of 30 and 42 we find their GCF, that is 6.

So to reduce $\dfrac{30}{42} = \dfrac{^{30}/_6}{^{42}/_6} = \dfrac{5}{7}$ We divide both the numerator and denominator by the GCF.

Example 3

Reduce $^{102}/_{40}$ to its lowest terms using the factorizing method.

Solution

$102 = 2 \times 3 \times 17$ $40 = 2 \times 2 \times 2 \times 5$

So, $\dfrac{102}{42} = \dfrac{2 \times 3 \times 17}{2 \times 3 \times 7} = \dfrac{51}{7}$

$\therefore\ \dfrac{102}{40} = \dfrac{51}{20}$ in lowest terms

Example 4

Reduce the fractions to their lowest terms by the GCF method;

(i) $\dfrac{100}{550}$ (ii) $\dfrac{108}{36}$

Solution

(i) F_{100} = { 1, 2, 4, 5, 10, 20, 25, 50, 100}

 F_{550} = { 1, 2, 5, 10, 11, 22, 25, 50, 55, 110, 275, 550}

∴ $\dfrac{100}{550}$ = $\dfrac{100/50}{550/50}$ = $\dfrac{2}{11}$

(ii) F_{108} = { 1, 2, 3, 4, 6, 9, 12, 18, 27, 36, 54, 108}

 F_{300} = { 1, 2, 3, 4, 5, 6, 10, 12, 15, 20, 25, 30, 50, 60, 75, 100, 150, 300}

∴ $\dfrac{108}{300}$ = $\dfrac{108/12}{300/12}$ = $\dfrac{9}{25}$

Activity 5.8

Reduce the fractions to their lowest terms (Use both methods in each case).

(i) $\dfrac{36}{144}$ (ii) $\dfrac{88}{108}$ (iii) $\dfrac{250}{55}$ (iv) $3\dfrac{11}{33}$

Always remember that when dealing with these two methods we involve canceling in both.

5.3   Comparing fractions

Given to compare the values of fractions $^2/_5$ and $^3/_5$, we can easily tell that $^2/_5$ is less than $^3/_5$. However it is very difficult to directly compare fractions such as $^5/_8$ and $^7/_9$.

For comparison of such fractions it requires first expressing them as decimals, percentages or as whole numbers, using the Lowest Common Multiple (LCM) of the denominators. In this case we are going to use the LCM of denominators, however use of decimals and percentages will be dealt with in the later chapters 6 and 7 respectively.

Example 5

Arrange the following fractions in ascending order;

(i) $\dfrac{2}{3}$, $\dfrac{3}{7}$, $\dfrac{8}{14}$, $\dfrac{5}{6}$, $\dfrac{1}{6}$ (ii) $\dfrac{2}{5}$, $\dfrac{3}{10}$, $\dfrac{5}{6}$, $\dfrac{1}{5}$, $\dfrac{2}{3}$

Solution

First get the LCM of the denominators. Then multiply each fraction with the LCM, the results obtained will be whole numbers easy to arrange in order of size.

(i) $\frac{2}{3}, \frac{3}{7}, \frac{8}{14}, \frac{5}{6}, \frac{1}{6}$

Find the LCM of denominators 3, 6, 7 and 14 which is 42.

Then: $\frac{2}{3} \times 42 = 28$; $\frac{3}{7} \times 42 = 18$; $\frac{9}{14} \times 42 = 27$;

$\frac{5}{6} \times 42 = 35$; $\frac{1}{6} \times 42 = 7$

$\therefore$ The arrangement in ascending order is; $\frac{1}{6}, \frac{3}{7}, \frac{9}{14}, \frac{2}{3}, \frac{5}{6}$

(ii) $\frac{2}{5}, \frac{3}{10}, \frac{5}{6}, \frac{1}{5}, \frac{2}{3}$ LCM of 3, 5, 6 and 10 is 30.

Then: $\frac{2}{5} \times 30 = 12$; $\frac{3}{10} \times 30 = 9$; $\frac{5}{6} \times 30 = 25$;

$\frac{1}{5} \times 30 = 6$; $\frac{2}{3} \times 30 = 20.$

$\therefore$ The arrangement in ascending order is; $\frac{1}{5}, \frac{3}{10}, \frac{2}{5}, \frac{2}{3}, \frac{5}{6}$

Exercise 5a

1 Express the following fractions as improper fractions

   (i) $^{27}/_{11}$ (ii) $^{132}/_{10}$ (iii) $^{17}/_{12}$ (iv) $2\,^{8}/_{11}$ (v) $3\,^{9}/_{10}$

2 Express the following fractions as mixed fractions

   (i) $\frac{21}{5}$ (ii) $\frac{63}{41}$ (iii) $\frac{32}{5}$ (iv) $\frac{11}{6}$ (v) $\frac{100}{23}$

3 Reduce the following fractions to their lowest terms.

   (i) $\frac{204}{16}$ (ii) $\frac{155}{20}$ (iii) $\frac{33}{99}$ (iv) $\frac{11}{341}$ (v) $\frac{120}{1000}$

4 Write at least 5 fractions in each case which are equivalent to the following.

(i) $\dfrac{3}{11}$ (ii) $\dfrac{8}{7}$ (iii) $\dfrac{4}{9}$ (iv) $\dfrac{31}{50}$ (v) $\dfrac{29}{30}$

5 Arrange the following fractions in descending order

(i) $\dfrac{2}{5}$, $\dfrac{1}{11}$, $\dfrac{2}{7}$, $\dfrac{6}{11}$, $\dfrac{3}{5}$

(ii) $\dfrac{4}{9}$, $\dfrac{1}{3}$, $\dfrac{3}{4}$, $\dfrac{5}{9}$, $\dfrac{4}{3}$

(iii) $\dfrac{2}{7}$, $\dfrac{11}{12}$, $\dfrac{4}{7}$, $\dfrac{5}{6}$, $\dfrac{1}{4}$

(iv) $\dfrac{3}{5}$, $\dfrac{1}{8}$, $\dfrac{3}{4}$, $\dfrac{7}{10}$, $\dfrac{5}{8}$

(v) $\dfrac{1}{2}$, $\dfrac{1}{4}$, $\dfrac{4}{11}$, $\dfrac{5}{9}$, $\dfrac{5}{12}$

5.4.1 Addition of fractions

To add fractions of similar denominators we can directly add the numerators and maintain the denominators.

Example 1

Add: (i) $\dfrac{1}{5} + \dfrac{3}{5}$ (ii) $\dfrac{2}{7} + \dfrac{4}{7}$

Solution

(i) $\dfrac{1}{5} + \dfrac{3}{5} = \dfrac{1+3}{5}$

$= {}^4/_5$

(ii) $\dfrac{2}{7} + \dfrac{4}{7} = \dfrac{2+4}{7}$

$= {}^6/_7$

If the denominators have a common term, make the denominators the same since one is a factor of the other. Remember equivalent fractions;

Example 2

Add: $\dfrac{3}{5} + \dfrac{1}{10}$

Solution

$\dfrac{3}{5} = \dfrac{3}{5} \times \dfrac{2}{2} = \dfrac{6}{10}$ Now the denominators are the same in both terms being added, that is, 10.

We now have: $\dfrac{3}{5} + \dfrac{1}{10} = \dfrac{6}{10} + \dfrac{1}{10} = \dfrac{6+1}{10} = \dfrac{7}{10}$

Example 3

Add: $\dfrac{4}{11} + \dfrac{1}{2}$

Solution

Make the denominators similar, then add the numerators maintaining the common denominator.

$$\dfrac{4}{11} + \dfrac{1}{2} = \dfrac{4}{11} \times \dfrac{2}{2} + \dfrac{1}{2} \times \dfrac{11}{11} = \dfrac{8}{22} + \dfrac{11}{22}$$

$$= \dfrac{19}{22}$$

Activity 5.9

In each of the following questions make the denominators equal using the technique of equivalent fractions, then carry out the additions;

(i) $\dfrac{2}{7} + \dfrac{9}{14}$ 　　(ii) $\dfrac{1}{5} + \dfrac{7}{15}$ 　　(iii) $\dfrac{3}{8} + \dfrac{2}{5}$ 　　(iv) $\dfrac{2}{7} + \dfrac{1}{11}$

In cases where the denominators are different for the terms being added and the common denominator cannot easily be got, then the LCM of the denominators is used.

Example 4

Add: (i) $\dfrac{8}{15} + \dfrac{5}{6}$ 　　　　(ii) $\dfrac{9}{4} + \dfrac{1}{7}$ 　　　　(iii) $2\dfrac{1}{3} + 1\dfrac{1}{4}$

Solution

(i) The LCM of 15 and 6 is 30

$$\dfrac{8}{15} + \dfrac{5}{6} = \dfrac{(2 \times 8) + (5 \times 5)}{30}$$

$$= \dfrac{16 + 25}{30}$$

$$= \dfrac{41}{30}$$

(or $1\,{}^{11}/_{30}$ as a mixed fraction)

(ii) The LCM of 4 and 7 is 28

$$\dfrac{9}{4} + \dfrac{1}{7} = \dfrac{(7 \times 9) + (4 \times 1)}{28}$$

$$= \dfrac{63 + 4}{28}$$

$$= \dfrac{67}{28}$$

(or $2\,{}^{11}/_{28}$ as a mixed fraction)

(iii) $2\,^1/_3 + 1\,^1/_4$ $= (2+1) + (\,^1/_3 + \,^1/_4)$

$= 3 + (\,^1/_3 + \,^1/_4)$ But the LCM of 3 and 4 is 12.

$$= 3\ \frac{[(4\times1)+(3\times1)]}{12}$$

$$= 3\ \frac{(4+3)}{12} \qquad = 3\,^7/_{12}$$

Activity 5.10

1 Work out numbers (i) in example 1 and (i) in example 2 using LCMs. Remember that the LCM of 5 and 5 is 5 and the LCM of 5 and 10 is 10.
2 Work out number (iii) in example 4 by first changing the terms into improper fractions, then use the LCM.

We can generally add fractions as follows especially where it is very difficult to find the LCM of the denominators .

Given to add: $\dfrac{a}{b} + \dfrac{c}{d}$ where a, b, c and d take any value but not zero.

we obtain, $\dfrac{a}{b} + \dfrac{c}{d} = \dfrac{a}{b} \bowtie \dfrac{c}{d} = \dfrac{ad + bc}{bd}$

This gives a general approach which is precise, but may involve harder multiplications. In this case the sign " $\leftrightarrow$ " denotes multiplication.

For example:

$\dfrac{3}{7} + \dfrac{9}{11} \Rightarrow \dfrac{3}{7} \bowtie \dfrac{9}{11} = \dfrac{(3\times11)+(9\times7)}{11} = \dfrac{33 + 63}{7\times11} = \dfrac{96}{77}$ (or $1\,^{19}/_{77}$)

Exercise 5b

Add the following fractions:

1 $\dfrac{5}{8} + \dfrac{1}{8}$ 2 $\dfrac{7}{11} + \dfrac{3}{22}$ 3 $\dfrac{9}{4} + \dfrac{4}{11}$ 4 $\dfrac{14}{21} + \dfrac{3}{11}$

5 $\dfrac{6}{17} + \dfrac{22}{23}$ 6 $\dfrac{31}{50} + \dfrac{1}{8} + \dfrac{7}{10}$ 7 $2\dfrac{3}{8} + 3\dfrac{9}{11}$ 8 $4\dfrac{1}{11} + \dfrac{3}{6} + 1$

9 $3\dfrac{22}{79} + 2\dfrac{8}{21}$ 10 $4\dfrac{2}{9} + 1\dfrac{1}{4} + 5\dfrac{3}{7}$

5.4.2 Subtraction of fractions

To subtract fractions we shall follow the same procedures as in addition, however, we need to replace the addition sign with the subtraction sign.

Example 1

Subtract: (i) $\dfrac{3}{8} - \dfrac{1}{8}$ (ii) $\dfrac{2}{3} - \dfrac{1}{6}$

Solution

(i) $\dfrac{3}{8} - \dfrac{1}{8} = \dfrac{3-1}{8}$

$= \dfrac{2}{8}$

$= \dfrac{1}{4}$

(ii) $\dfrac{2}{3} - \dfrac{1}{6} = \dfrac{2}{3} \times \dfrac{2}{2} - \dfrac{1}{6}$

$= \dfrac{4}{6} - \dfrac{1}{6}$

$= \dfrac{3}{6} = \dfrac{1}{2}$

Example 2

Subtract: (i) $4\,^7/_8 - 2\,^2/_5$ (ii) $8 - \,^5/_7$

Solution

(i) The LCM of 5 and 7 is 35

$\dfrac{8}{5} - \dfrac{4}{7} = \dfrac{(7 \times 8) - (5 \times 4)}{35}$

$= \dfrac{56 - 20}{35}$

$= \dfrac{36}{35}$

$= 1\,^1/_{35}$

(ii) $4\,^7/_8 - 2\,^2/_5 = (4 - 2) + (^7/_8 - \,^2/_5)$

LCM of 8 and 5 is 40.

$\dfrac{2\,[(5 \times 7) - (8 \times 2)]}{40}$

$= \dfrac{2\,(35 - 16)}{40}$

$= 2\,^{19}/_{40}$

Example 3

Subtract: $2\,^1/_{10} - \,^6/_{13}$

Solution

$2\,^1/_{10} - \,^6/_{13} = \dfrac{21}{10} - \dfrac{6}{13} \Rightarrow \dfrac{21}{10} - \dfrac{6}{13} = \dfrac{(21 \times 13) - (10 \times 6)}{10 \times 13} = \dfrac{273 - 60}{130} = \dfrac{213}{130}$

Exercise 5c

Subtract the following fractions

1 $\dfrac{7}{11} - \dfrac{2}{11}$ 2 $\dfrac{3}{10} - \dfrac{3}{20}$ 3 $\dfrac{4}{9} - \dfrac{2}{11}$ 4 $\dfrac{9}{17} - \dfrac{1}{7}$

5 $\dfrac{9}{4} - 1\dfrac{1}{4}$ 6 $\dfrac{10}{7} - \dfrac{3}{10}$ 7 $\dfrac{33}{40} - \dfrac{20}{53}$ 8 $21\dfrac{1}{8} - 14\dfrac{4}{11}$

9 $3 - \dfrac{7}{11}$ 10 $1\dfrac{1}{2} - 1 - \dfrac{3}{8}$

5.4.3 Multiplication of fractions

Multiplication by use of models

¼ can be represented as;   and   ¹/₃ can be represented as;

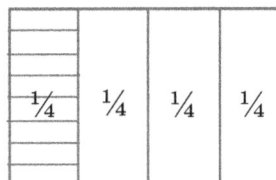   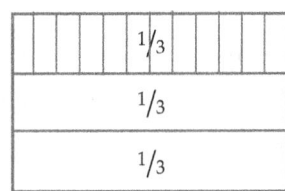

Then ¼ x ¹/₃ ⇒

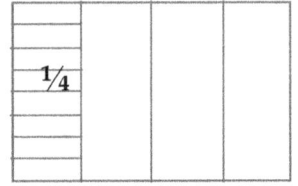   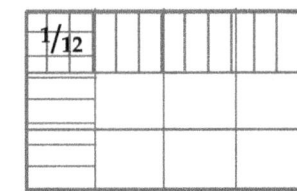

We have the representative for ¼ put exactly onto the representative of ¹/₃. We notice that in the answer there are 12 portions. So to obtain the fraction in the answer, we take the number of portions shaded twice over the total number of portions.

Similarly; ²/₅ x ¼ ⇒ ²/₅ x ¼ = ²/₂₀ = ¹/₁₀ (in lowest terms)

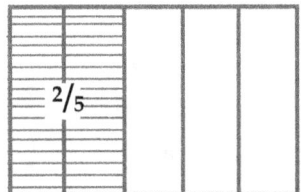

Activity 5.11

Work out the following diagrammatically: (i) $^2/_7 \times {}^1/_6$ (ii) $^2/_3 \times {}^5/_8$

Example 1

Multiply: (i) $^{29}/_{30} \times {}^{15}/_{13}$ (ii) $3\,^1/_5 \times 2\,^1/_2$

Solution

(i) $^{29}/_{30} \times {}^{15}/_{13} = \dfrac{29 \times 15}{30 \times 13} = \dfrac{29 \times 1}{2 \times 13}$

$= \dfrac{29}{26}$

$= 1\,^3/_{26}$

(iii) $3\,^1/_5 \times 2\,^1/_2 = \dfrac{\overset{8}{16}}{5} \times \dfrac{5}{\underset{1}{2}}$

$= \dfrac{8 \times 1}{1 \times 1}$

$= 8$

While multiplying fractions always remember to;
1. To change the mixed fractions to improper ones first then proceed.
2. Cancel any common factors existing in both the numerator and the denominator.

Exercise 5d

Work out:

1. $8 \times \dfrac{7}{32}$

2. $\dfrac{1}{3} \times \dfrac{9}{7}$

3. $\dfrac{2}{5} \times \dfrac{15}{22}$

4. $\dfrac{8}{11} \times \dfrac{12}{100}$

5. $2\dfrac{1}{8} \times 3\dfrac{5}{17}$

6. $3\dfrac{2}{9} \times \dfrac{23}{76}$

7. $8\dfrac{3}{11} \times 1\dfrac{10}{23}$

8. $3\dfrac{4}{9} \times 10\dfrac{4}{5} \times \dfrac{5}{31}$

9. $33\dfrac{1}{8} \times \dfrac{11}{265} \times \dfrac{1}{121}$

10. $\dfrac{50}{31} \times \dfrac{62}{15} \times \dfrac{3}{5}$

5.4.4 Division of fractions

If we you are given $12 \div 2$, it means how many 2s are in 12? Similarly, $8 \div {}^1/_5$ means how many $^1/_5$s are in 8. Dividing by a number means multiplying by

---

its reciprocal, for example;

$8 \div 4 = 8 \times \frac{1}{4}$, since the reciprocal of 4 is $\frac{1}{4}$.

So, $8 \div 4 = 8 \times \frac{1}{4} = 2$, and

$\frac{2}{5} \div \frac{4}{3} = \frac{2}{5} \times \frac{3}{4}$, since the reciprocal of $\frac{4}{3}$ is $\frac{3}{4}$.

Example 1

Divide: (i) $\frac{3}{8} \div \frac{1}{4}$ (ii) $1\frac{5}{8} \div \frac{13}{11}$

Solution

(i) $\frac{3}{4} \div \frac{1}{4}$ $= \frac{3}{8} \times \frac{4}{1}$ (ii) $1\frac{5}{8} \div \frac{13}{11}$ $= \frac{13}{8} \div \frac{13}{11}$

$= \frac{3 \times 4}{8 \times 1}$ $= \frac{13}{8} \times \frac{11}{13}$

$= \frac{3 \times 1}{2 \times 1}$ $= \frac{11}{8}$

$= \frac{3}{2}$ $= 1\frac{3}{8}$

Remember that;

(i) *'Dividing by a number'* equals J*multiplying by its reciprocal'* , for example, a number, m, divided by n is equal to the number, m, multiplied by the reciprocal of n, that is; $(m \div n) = (m \times \frac{1}{n})$.

(ii) The reciprocal of a number n is $\frac{1}{n}$, that is, the reciprocal of 3 is $\frac{1}{3}$.

(iii) The reciprocal of a fraction $\frac{a}{b}$ is $\frac{1}{\frac{a}{b}}$, which is equal to $\frac{b}{a}$.

The reciprocal of $\frac{2}{3}$ is $\frac{1}{\frac{2}{3}}$ $= \frac{3}{2}$

Activity 5.12

Work out :   (i) $\frac{3}{4} \div 6$ (ii) $2 \div \frac{1}{7}$ (iii) $\frac{1}{8} \div 8$ (iv) $5 \div \frac{1}{5}$

Exercise 5e

1   $\frac{4}{7} \div \frac{21}{28}$ 2 $\frac{3}{7} \div \frac{63}{11}$ 3 $\frac{88}{111} \div \frac{11}{2}$

4 $2\frac{4}{11} \div \frac{33}{104}$ 5 $\frac{111}{8} \div \frac{89}{8}$ 6 $38\frac{2}{3} \div 3\frac{2}{9}$

7 $4\frac{11}{21} \div \frac{15}{42}$ 8 $\frac{9}{4} \div \frac{81}{400}$

9 $1\frac{23}{121} \div \frac{12}{33}$ 10 $23\frac{9}{11} \div 3\frac{11}{40}$

5.4.5 Mixed operations

When we are given to workout fractions with more than one sign, we resort to BODMAS.

Example 1

Work out: (i) $(^2/_3 - ^1/_5) + ^3/_2 \div ^9/_4$ of $^1/_3$ (ii) $^2/_{11} + ^3/_5 \div ^6/_7$ of $^1/_3$

Solution

(i) $(^2/_3 - ^1/_5) + ^3/_2 \div ^9/_4$ of $^1/_3$

$\frac{(10-3)}{15} + \frac{3}{2} \div \frac{9}{4}$ of $\frac{1}{3}$

$\frac{7}{15} + \frac{3}{2} \div \frac{9}{4} \times \frac{1}{3}$

$\frac{7}{15} + \frac{3}{2} \div \frac{3}{4}$

$\frac{7}{15} + \frac{3}{2} \times \frac{4}{3}$

$\frac{7}{15} + \frac{2}{1}$

$\frac{7+30}{15} = \frac{37}{15}$

$= 2\,^7/_{15}$

(ii) $^2/_{11} + ^3/_5 \div ^6/_7$ of $^1/_3$

$^2/_{11} + ^3/_5 \div \frac{6}{7} \times \frac{1}{3}$

$^2/_{11} + ^3/_5 \div \; ^2/_7$

$^2/_{11} + (^3/_5 \div \; ^2/_7)$

$^2/_{11} + \frac{3}{5} \times \frac{7}{2}$

$\frac{2}{11} + \frac{21}{10}$

$\frac{20+231}{110} = \frac{251}{110}$

$= 2\,^{31}/_{110}$

Exercise 5f

Evaluate the following fractions.

1 $\frac{3}{8} + \frac{2}{5} \div \frac{4}{3} - \frac{1}{3} + \left(\frac{3}{11}$ of $\frac{33}{12}\right)$

2 $\frac{2}{11}$ of $24\frac{1}{5} - \left(\frac{11}{13} + \frac{7}{5}\right)$

3 $\left(\frac{2}{3} - \frac{1}{8}\right) \div 1\frac{3}{4} - \frac{7}{8}$

4 $\frac{7}{10} \div \frac{21}{20} + \frac{3}{8} - \left(\frac{1}{2} \times \frac{3}{5}\right)$

5   $\dfrac{(\frac{1}{2}+\frac{3}{5}-\frac{1}{3}) \times 1\,^{7}/_{23}}{(^{3}/_{5}-\,^{1}/_{8}) \div \,^{11}/_{40}}$

6   $\left[\dfrac{30}{51} \text{ of } \dfrac{102}{15}\right] \div \dfrac{1}{4} + \dfrac{19}{17} - \dfrac{2}{17}$

7   $4\dfrac{1}{9} - 2 + \dfrac{6}{11} \div \dfrac{3}{22}$

8   $6\dfrac{2}{5} \div \dfrac{4}{5} + \dfrac{19}{17} - \dfrac{2}{17}$

9   $18 \times \dfrac{5}{9} - \dfrac{6}{7} + \dfrac{3}{10} \div \dfrac{7}{20}$

10   $\dfrac{3}{7} + \dfrac{9}{11} \div \left[\dfrac{12}{11} \text{ of } 2\dfrac{1}{7}\right]$

11   $\dfrac{^{2}/_{9} + \,^{3}/_{5} - \,^{1}/_{8}}{^{1}/_{44} \div \,^{90}/_{11}}$

12   $\left[4\dfrac{2}{3} - 3\dfrac{5}{11}\right] \times \left[2\dfrac{1}{4} + 1\dfrac{1}{3} - \dfrac{5}{6}\right]$

13   $\dfrac{5\,^{1}/_{3} + 4\,^{2}/_{5}}{4\,^{13}/_{15}}$

14   $\dfrac{4\,^{1}/_{2} - \,^{15}/_{4} \times \,^{2}/_{9}}{^{12}/_{139} \times (\,^{13}/_{5} + \,^{7}/_{4} \div \,^{14}/_{7})}$

15   $\dfrac{(^{2}/_{3} + \,^{2}/_{5} - \,^{1}/_{2}) \div \,^{17}/_{45}}{2\,^{2}/_{5} \times (^{1}/_{3} - \,^{1}/_{8})}$

5.4.6 Problems involving fractions

If you go to the market to buy some vegetables you may use fractions of the total amount of money you have to allocate each of the vegetable type, say onions for Sh.200, tomatoes for Sh.350, green pepper for Sh.150, and so on. If you had a total amount of Sh.2000, then you will have used $^{1}/_{10}$ of it to buy onions.

Example 1

I save ¼ of my monthly salary, spend ½ of the rest on food and the remaining is spent on clothing. If I earn Sh.40,000 per month. Find how much money.
(i) I save.   (ii) I spend on food   (iii) I spend on clothing.

Solution

(i) Money saved
    = Sh.(¼ x 40,000)
    = Sh.10,000.

(ii) Money spent on food
    = Sh.[½(40,000 – 10,000)]
    = Sh.[½ x 30,000]
    = Sh.15,000

(iii) Money spent on clothing   = Sh.[40,000 – (10,000 + 15,000)]
    = Sh.(40,000 – 25,000)
    = Sh.15,000

Exercise 5g

1 A man is four and half times as old as his son. If the son is 10 years old, how old is the man?

2 In a school of 550 students; $1/5$ of them had brown shoes, the number of those with black shoes was two and half times the number of those with brown shoes. The rest of the students had grey shoes. Find how many students had, (a) brown shoes   (b) black shoes   (c) grey shoes ?

3 In playing a certain game Kato earned the following marks.

Stage 1	*Stage 2*
1st score = $1/8$ of a point	1st score = $-3/4$ of a point
2nd score = 1 point	2nd score = $5/8$ of point
3rd score = 4 point	3rd score = -2 points
4th score = $2/5$ of point	4th score = 1/4 of point

Stage 3 (*Final stage*)

1st score = $1/8$ of 3rd score in stage 1

2nd score = $3/4$ of 4th score in stage 1

3rd score = $3/5$ of the 2nd score stage 2

4th score = 4/5 of the 3rd score stage 2.

   (a) Find the total number of points scored by Kato in each stage of the game.
   (b) What was the total number of points he gathered in the 3 stages?
   (c) If the total number of points gathered in all the 3 stages is grater than two and half times of the points gathered in stage two then Kato wins the game . Did Kato win this game?

4 A book has 640 pages, $4/5$ of these pages are written on, $1/8$ of them have figures, $1/128$ of them are blank and the rest are written on but half full.
   Find:- (a) How many pages are
                (i) written on   (ii) have figures   (iii) blank. (iv) half full.
            (b) The number of half full pages as a fraction of the total number of the pages.

5 A quarter of the boys and third of the girls in a class of 48 got the answer correct. If three quarters of the class are boys;

(a) How many boys and girls are there in the class?

(b) How many boys;

    (i) got the answer correct?   (ii) failed the answer correct?

(c) How many girls; (i) got the answer correct ?   (ii) failed the answer?

(d) What fraction of the girls failed the answer ?

(e) Express the number of the boys who passed the question as a fraction of the whole class.

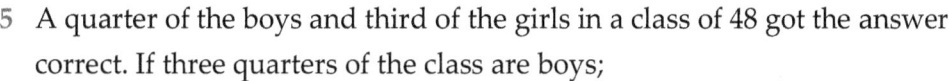

REVISION EXERCISE 5

1 (a) Write the fractions represented by the following figures:

    (i) 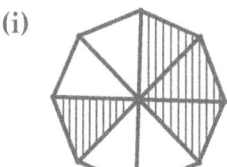   (ii)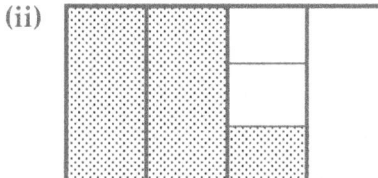

(b) Change the following fractions to improper fractions

    (i) $3\frac{1}{7}$ (ii) $14\frac{2}{9}$

(c) Change the following fractions to mixed fractions (i) $\frac{111}{50}$ (ii) $\frac{63}{22}$

2 (a) Reduce the following to lowest terms   (i) $\frac{801}{900}$ (ii) $\frac{3146}{5049}$

(b) Write two equivalent fraction in each case to: (i) $\frac{4}{11}$ (ii) $\frac{13}{20}$

(c) Arrange in ascending order:

    (i) $\frac{1}{2}, \frac{3}{5}, \frac{4}{7}, \frac{2}{11}, \frac{3}{2}$ (ii) $\frac{3}{11}, \frac{13}{20}, \frac{8}{11}, \frac{2}{7}, \frac{5}{8}$

3 Represent the following shaded parts as fractions and carry out the multiplications using these figures and state your answer in their lowest terms.

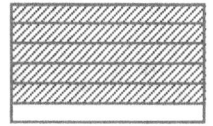 x  = 

Work out the following:

4   (a)   (i) $2/5 + 1\,3/8 + 2\,1/2 + 1$ (ii) $5\,1/7 + 1/2 + 1\,1/4 + 8/11$

       (b)   (i) $3\,1/3 - 1/2 - 1\,3/7 - 10/11$ (ii) $5 - 1\,1/2 - 2\,6/7 - 1/6$

5   (a)   (i) $1\,1/2 \times 4/7 \times 1\,5/9$ (ii) $8/9 \times 2\,1/4 \times 1/2$

       (b)   (ii) $1/3 \div 9/10$ (ii) $2\,1/3 \div 9/26$

6   (i) $1\,\tfrac{1}{2} + (3\,\tfrac{1}{4} - 2\,1/3)$ of $1\,1/11$ (ii) $1\,1/3 \div 2/9 + 1/3 - (3/5$ of $10/9)$

7 In a certain test students either passed, in grade 1, grade 11 or failed. $3/8$ of the boys in the class failed and no boy passed in grade 1. $\tfrac{3}{4}$ of the girls passed in grade 1, $1/8$ of the girls passed in grade 11 and the rest failed. If there were 80 students in the class and $3/5$ of the class were boys.
   Find: (a) (i) The number of boys in class (ii) The number of girls in class.

           (b) The number of students who,   (i)   Passed in grade 1

                                      (ii) Passed in grade 11

                                      (iii) Failed

       (c)   What fraction of the whole class passed?

       (d)   What was the fraction of,   (i) Girls who passed?

                                        (ii) Boys who passed?

       (e)   Of the girls and boys, who generally performed better than the others?

8   (a)   27 cubes of equal volume were pilled and glued to form a single larger cube. The large cube was then painted on all its surfaces. Find how many small cubes had, and express them as fractions (in lowest terms) of the total number of small cubes.

           (i)   None of their faces painted   (ii)   1 face painted

           (iii)   2 faces painted   (iv)   3 faces painted.

           (v)   4 faces painted;

       (b)   A watch loses $3/5$ of a second every week.

           (i) Find the number of seconds the watch will have lost after 10 weeks?

           (ii) What fraction of a minute will this be?

           (iii) If you set this watch to read 11:30 am, what will the actual time be if it reads 11:30 pm after 100 weeks?

DECIMAL FRACTIONS

INTRODUCTION

Earlier, we looked at numbers and common fractions respectively. We realize that whole numbers are exact, say 110, 2, 41, 9, etc.; where as fractions are smaller denominations of whole number 1. We can have common fractions of 1, as ½ of 1, $^1/_3$ of 1, $^2/_5$ of 1, etc.

Decimal fractions are another representation of fractions that involve a decimal point separating the whole number part from the decimal part, that is; the fraction ½ can be represented as 0.5 in decimal form; $^4/_5$ as 0.8 in; 1 ½ as 1.5, etc.

6.1 Extending place values to decimals

If we have a decimal say 139.4936, we can describe it using a place value chart as shown below, where its digits take the following place values;

T	O	.	Tth	Hth	Thth	TThth	HThth
3	9	.	4	9	3	6	

Whole number part   D.P   Decimal number part

From the d.p the values increase towards the left and decrease towards the right. From the chart we can then have:

T	(10^1)	$\Rightarrow$ 3 *Tens* (3 x 10)	3 0
O	(10^0)	$\Rightarrow$ 9 *Ones* (9 x 1)	9
Tth	(10^{-1})	$\Rightarrow$ 4 *Tenths* (4 x 0.1)	0 . 4
Hth	(10^{-2})	$\Rightarrow$ 3 *Hundreds* (9 x 0.01)	0 . 0 9
Thth	(10^{-3})	$\Rightarrow$ 6 *Thousandths* (3 x 0.001)	0 . 0 0 3
TThth	(10^{-4})	$\Rightarrow$ 3 *Tens of Thousandths* (6 x 0.0001)	+ 0 . 0 0 0 6
			3 9 . 4 9 3 6

Remember: Tth means Tenths ($^1/_{10}$), Hth means Hundredths ($^1/_{100}$), and so on.

Exercise 6a

Write the following fractions as decimal numbers:

1 $^1/_{10}$ 2 $^3/_{100}$ 3 $^9/_{1000}$ 4 $^{17}/_{100}$ 5 $^{43}/_{10000}$

What is the place value of 2 in each of the following:

6 24.61   7 4469.30121   8 14167.2001   9 304.33812   10 0.3396214

If 341.32 is expanded as:

$(3 \times 100 + 4 \times 10 + 1 \times 1 + 3 \times {}^1/_{10} + 2 \times {}^1/_{100}) = (300 + 40 + 1 + 0.3 + 0.02)$; then expand the following:

11 1316.41   12 401.617   13 60.4116   14 0.3041   15 0.00214

6.1.1 Comparing decimals

Different decimals have different values depending on the digitsJ and their places. The places increase value from the decimal point towards the left and decrease value from the decimal point towards the right as earlier stated.

To compare 0.9721 and 0.961 we start from the left hand side, checking the digits of similar place value.

O	.	Tth	Hth	Thth	TThth
0	.	9	7	2	1
0	.	9	6	1	0

For the Ones and Tens, the digits are the same. So we can not judge using them.

In the column of Hundreds, we have 7 in the first number, that is, 7 hundredths ($^7/_{100}$) and the second, 6 hundredths ($^6/_{100}$).
But $^7/_{100} > {}^6/_{100}$; therefore 0.9721 > 0.961.

Generally look for the decimal number with least whole number value then followed by those values which have lower digits as you move from the decimal point to the right, to an arrangement from the smallest to highest value (ascending order).

Example 1

Arrange the following decimals in ascending order: 0.412, 4.120, 0.0412, 12.4

Solution

H	T	O		Tth	Hth	Thth	TThth	Ascending order
		0	.	4	1	2		2nd
		4	.	1	2	0		3rd
		0	.	0	4	1	2	1st
	1	2	.	4				4th

We therefore have the arrangement in ascending order as:

0.0412, 0.412, 4.120, 12.4

Example 2

Arrange in ascending order: 13.32, 1.332, 133.2, 0.1332

Solution

H	T	O		Tth	Hth	Thth	TThth	Ascending Order
	1	3	.	3	2			3rd
		1	.	3	3	2		2nd
1	3	3	.	2				4th
		0	.	1	3	3	2	1st

Now the arrangement in ascending order is: **0.1332, 1.332, 13.32, 133.2**

Exercise 6b

Arrange the following decimals in ascending order.

1 0.0412, 0.412, 4.12, 0.00412, 4.012

2 14.101, 1.4101, 141.01, 0.14101, 0.0014101

3 123.11, 12.311, 0.0012311, 1.2311, 1231.1

4 34.12, 121.2, 5.21, 21.5, 41.51

5 11.1, 10.1, 0.101, 0.0010, 0.0011.

6 0.12301, 12.301, 2.30101, 3.03201, 0.01231

7 8.00808, 0.080880, 0.080808, 0.8880808

8 0.001020304, 0.00001234, 0.10001234

6.2 Addition and Subtraction

To add or subtract two decimal numbers you arrange the numbers one over the other such that the decimal point lies on the same vertical line, then carry out the addition and subtraction.

Example 1

Work out the following: (i) 41.23 + 91.101 + 401.14   (ii) 342.47 − 39.118

Solution

Rearrange the decimal numbers such that decimal points lie on the same vertical line.

```
(i)   41.230              (ii)
      91.101                    342.470
  + 401.140                 −    39.118
    533.471                     303.352
```

Activity 6.1

Work out: (i) 240 + 3.48 + 0.001   (ii) 41.8 − 2.189   (iii) 4 − 2.451

Exercise 6c

Work out the following:

1   13.42 + 34.112   2   36.21 + 3.6001   3   73.21 + 85.1001

4   0.0014 + 3   5   6.741 + 0.01002   6   3 04 − 0.0004

7   6.04 − 3.61   8   4.421 − 3.12   9   3.41 − 0.0142

10 13.42 − 1.40009

6.3 Multiplication and division of decimals

Here we carry out multiplication as done on whole numbers, but the sum of decimal places in the *multiplier* and *multiplicand* should be equal to the number of decimal places in the *answer*.

Assume no decimal places in your arrangement, however do not forget where to put the decimal point in your answer.

Example 1

Work out: (i) 24.32 x 38   (ii) 3.41 x 4.21

Solution

(i)
$$
\begin{array}{r}
2\,4\,.\,3\,2 \\
\times\ \ 3\,8 \\
\hline
1\,9\,4\,.\,5\,6 \\
+\ 7\,2\,9\,.\,6\,0 \\
\hline
\mathbf{9\,2\,4\,.\,1\,6}
\end{array}
$$

∴ 24.32 x 38 = **924.56**

(ii)
$$
\begin{array}{r}
3\,.\,4\,1 \\
\times\ 4.\,2\,1 \\
\hline
3\,4\,1 \\
6\,8\,2\,0 \\
+1\,3\,6\,4\,0\,0 \\
\hline
\mathbf{1\,4\,3\,5\,6\,1}
\end{array}
$$

∴ 3.41 x 4.21 = **14.3561**

Example 2

Work out: (i) 32.48 ÷ 4   (ii) 77.77 ÷ 0.7

Solution

(i)
$$
\begin{array}{r}
\mathbf{8\,.\,1\,2} \\
4\,\overline{\big)\,3\,2\,.\,4\,8} \\
-\ 3\,2\ \ \ \ \\
\hline
4\ \ \\
-\ 4\ \ \\
\hline
8 \\
-\ 8 \\
\hline
\end{array}
$$

4 goes into 32, 8 times,
4 goes into 4, once
4 goes into 8, twice
Ensure that the decimal point in the answer lies on the same vertical line with that in the dividend.

(ii) $77.77 \div 0.7 = \dfrac{77.77}{0.7} \times \dfrac{10}{10}$

$$= \dfrac{777.7}{7}$$

$$
\begin{array}{r}
1\,1\,1\,.\,1 \\
7\,\overline{\big)\,7\,7\,7\,.\,7} \\
-\ 7\,7\,7\,.\,7 \\
\hline
-\ -\ -\ -
\end{array}
$$

Make the denominator a whole number by multiplying with a power with the number of zeros equal to the decimal places.

Remember that; $^{10}/_{10} = 1$, and multiplying by 1 does not affect the value of the number.

∴ 77.77 ÷ 0.7 = **111.1**

Exercise 6d

Work out the following:

1	3 x 0.72	2	8.67 x 2.1	3	18.21 x 3.67
4	4.12 x 67.03	5	6.718 x 5.002	6	4.26 ÷ 10
7	32.48 ÷ 4	8	93.34 ÷ 2	9	42.14 ÷ 3
10	0.72 ÷ 0.9	11	3.12 ÷ 0.3	12	0.06 ÷ 0.03
13	400 ÷ 1.2	14	640 ÷ 2.4	15	0.00216 ÷ 0.36

6.4   Mixed Operations on decimals

Here we are to apply BODMAS as earlier studied for the various operation in a particular question given.

Example 1

Work out: (i) $2.41 \times 3.4 \div 0.17 + 3.14$ (ii) $43.232 \div 0.4 - 4.2 + 31 \times 3.01$

Solution

(i) $2.41 \times (3.4 \div 0.17) + 3.14$

$2.41 \times \dfrac{3.4}{0.17} \times \dfrac{100}{100} + 3.14$

$2.14 \times \dfrac{340}{17} + 3.14$

$2.14 \times 20 \qquad + 3.14$

$(2.14 \times 20) \qquad + 3.14$

$\qquad\qquad 42.8 \quad + 3.14$

$\qquad\qquad\qquad = \mathbf{45.94}$

(ii) $43.232 \div 0.4 - 4.2 + 31 \times 3.01$

$\dfrac{432.32}{4} \quad - 4.2 + 31 \times 3.01$

$108.08 \quad - 4.2 + (31 \times 3.01)$

$108.08 \quad - 4.2 + 93.31$

$108.08 + 93.31 \quad - 4.2$

$\qquad 201.39 \quad - 4.2$

$\qquad\qquad = \mathbf{197.19}$

Exercise 6e

Work out the following

1   $24.21 \times 3 + 1.47 - 3$ of 4.21

2   $\dfrac{(3.412 + 0.381 - 3.01) \text{ of } 32.01}{13.428 \div 4.0}$

3 $\dfrac{12 \times 13.2 - 2.3 \text{ of } 18.04}{34.1}$

4 $\dfrac{132.88 \div 33 - 0.801 + 6}{3.14 \text{ of } 2.71}$

5 $\dfrac{(32 - 2.4 + 6) \text{ of } (12.4 + 6.1 - 8.2)}{(33.44 \div 1.1) + 3.46}$

6.5   Conversion of decimals to fractions

At the beginning of this chapter we saw that decimals are another representation of fractions, therefore we can change decimals to fractions.

Given a value 0.412 it means we have;

0 ones   →   No whole number

4 tenths   → $4 \times \dfrac{1}{10} = \dfrac{4}{10}$

1 hundredths   → $1 \times \dfrac{1}{100} = \dfrac{1}{100}$

2 thousandths   → $2 \times \dfrac{1}{1000} = \dfrac{2}{1000}$

This means: $0.142 = 0 + \dfrac{4}{10} + \dfrac{1}{100} + \dfrac{2}{1000}$, making the denominators 1000 throughout;

that is, $0 + \dfrac{4 \times 100}{10 \times 100} + \dfrac{1 \times 10}{100 \times 10} + \dfrac{2}{1000}$

$0 + \dfrac{400}{1000} + \dfrac{10}{1000} + \dfrac{2}{1000} = \dfrac{400 + 10 + 2}{1000} = \dfrac{412}{1000}$

So, $0.412 = \dfrac{412}{1000}$, and as a fraction, in lowest terms; $0.412 = \dfrac{103}{250}$

Exercise 6f

Express the following decimals as fractions.

1 0.031	2 31.21	3 0.0041	4 0.101	5 4.1102
6 0.41	7 0. 41	8 0.08	9 0.33	10 0.142
11 0.088	12 0.630	13 5.404	14 64.06	15 80.08

6.6　Conversion of fractions to decimals

6.6.1　Terminating fractions

(Decimals with a definite number of decimal places)

Example 1

Change the following fractions to decimals. (i) $^3/_8$　(ii) $^{19}/_4$

Solution

(i) $^3/_8 \Rightarrow$

```
        0 . 3  7  5
    8 | 3 . 0  0  0
       -0
        3  0
      - 2  4
         6  0
       - 5  6
          4  0
        - 4  0
          -  -
```

$\therefore \ ^3/_8 = \mathbf{0.375}$

(ii) $^{19}/_4 \Rightarrow$

```
          4 . 7 5
    4 | 1 9 . 0 0
       -1 6
         3 0
       - 2 8
          2 0
        - 2 0
          -  -
```

$\therefore \ ^{19}/_4 = \mathbf{4.75}$

Exercise　6g

Change the following fractions to decimals,

1 $\dfrac{3}{10}$	2 $\dfrac{21}{20}$	3 $\dfrac{23}{50}$	4 $\dfrac{141}{100}$	5 $\dfrac{7}{8}$
6 $\dfrac{11}{20}$	7 $\dfrac{3}{4}$	8 $\dfrac{21}{50}$	9 $\dfrac{5}{16}$	10 $\dfrac{7}{16}$

6.6.2　Conversion of fractions to recurring decimals

These are decimals with a repeated digit or group of digits to form an endless number of decimal places. Recurring decimals can be represented in various forms, that is;

(a) $0.222... \Rightarrow 0.\dot{2}$　or　$0.\bar{2}$　;

(b) $0.3666... \Rightarrow 0.3\dot{6}$　or　$0.3\bar{6}$　;

(c) $0.990099009900 ... \Rightarrow 0.\dot{9}90\dot{0}$　or　$0.\overline{9900}$

Example 1

Change the following decimals to fractions, (i) $^2/_9$ (ii) $^4/_{11}$

Solution

(i)

$$^2/_9 \Rightarrow 9 \overline{) \begin{array}{l} 0.2\ 2\ 2\ \ldots \\ 2\ .\ 0\ 0\ 0 \end{array}}$$

$$\begin{array}{r} -\,0 \\ \hline 2\ 0 \\ -\,1\ 8 \\ \hline 2\ 0 \\ -\,1\ 8 \\ \hline 2\ 0 \\ -\,1\ 8 \\ \hline 2 \end{array}$$

The digit 2 is repeated continuously, hence recurring.

$$\therefore \ ^2/_9 = \mathbf{0.222 \ldots}$$

(ii)

$$^4/_{11} \Rightarrow 11 \overline{) \begin{array}{l} 0.3\ 6\ 3\ 6\ 3\ 6\ \ldots \\ 4\ .\ 0\ 0\ 0\ 0\ 0\ 0 \end{array}}$$

$$\begin{array}{r} -\,0 \\ \hline 4\ 0 \\ -\,3\ 3 \\ \hline 7\ 0 \\ -\,6\ 6 \\ \hline 4\ 0 \\ -\,3\ 3 \\ \hline 7\ 0 \\ -\,6\ 6 \\ \hline 4\ 0 \\ -\,3\ 3 \\ \hline 7\ 0 \\ -\,6\ 6 \\ \hline 4 \end{array}$$

$$\therefore \ ^4/_{11} = \mathbf{0.363636 \ldots}$$

Example 2

Change the following fractions to decimals (i) $^7/_{30}$ (ii) $^{100}/_{101}$

(i)

$$^7/_{30} \Rightarrow 30 \overline{) \begin{array}{l} 0.2\ 3\ 3\ 3\ \ldots \\ 7\ .\ 0\ 0\ 0\ 0 \end{array}}$$

$$\begin{array}{r} -\,0 \\ \hline 7\ 0 \\ -\,6\ 0 \\ \hline 1\ 0\ 0 \\ -\,9\ 0 \\ \hline 1\ 0\ 0 \\ -\,9\ 0 \\ \hline 1\ 0\ 0 \\ -\,9\ 0 \\ \hline 1\ 0 \end{array}$$

Therefore, $^7/_{30} = \mathbf{0.2333 \ldots}$

(ii)

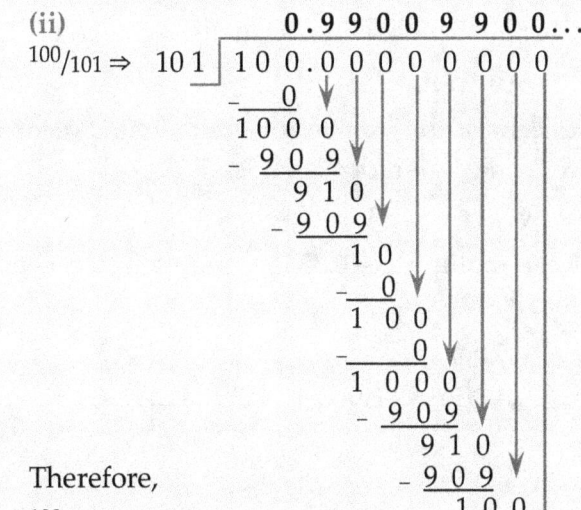

$$^{100}/_{101} \Rightarrow 101 \overline{) \begin{array}{l} 0.9\ 9\ 0\ 0\ 9\ 9\ 0\ 0\ldots \\ 1\ 0\ 0\ .\ 0\ 0\ 0\ 0\ 0\ 0\ 0\ 0 \end{array}}$$

$$\begin{array}{r} -\,0 \\ \hline 1\ 0\ 0\ 0 \\ -\,9\ 0\ 9 \\ \hline 9\ 1\ 0 \\ -\,9\ 0\ 9 \\ \hline 1\ 0 \\ -\,0 \\ \hline 1\ 0\ 0 \\ -\,0 \\ \hline 1\ 0\ 0\ 0 \\ -\,9\ 0\ 9 \\ \hline 9\ 1\ 0 \\ -\,9\ 0\ 9 \\ \hline 1\ 0\ 0 \\ -\,0 \\ \hline 1\ 0\ 0\ 0 \end{array}$$

Therefore,

$^{100}/_{101} = \mathbf{0.9900 \ldots}$

Exercise 6h

Change the following fractions to decimals.

1 $\dfrac{4}{15}$	2 $\dfrac{8}{9}$	3 $\dfrac{2}{5}$	4 $\dfrac{9}{11}$	5 $\dfrac{8}{11}$
6 $\dfrac{11}{12}$	7 $\dfrac{17}{30}$	8 $\dfrac{10}{101}$	9 $\dfrac{6}{13}$	10. $\dfrac{3}{7}$
11 $\dfrac{99}{101}$	12 $\dfrac{16}{21}$	13 $\dfrac{2}{13}$	14 $\dfrac{11}{90}$	15 $\dfrac{97}{30}$

Activity 6.2

Change the following fractions to decimals and comment on your answers.

(i) $5/17$ (ii) $4/9$ (iii) $10/17$ (iv) $11/35$

6.6.3 Conversion of recurring decimals to fractions

Given a recurring decimal such as; 0.333 ... , we have one recurring digit, 3. For 0.272727 ... , 2 digits are recurring; whereas in 0.34111 ... , 1 digit is recurring.

The technique used to convert such decimals to fractions will depend on the number of recurring digits and how they recur.

Example 1

Change the following decimals to fractions 0.333 ...

Solution

Check for the number of digits which recur in the decimal, 0.333 ... . In this case one digit recurs, that is, 3.

Now let $v = 0.333 \ldots$

Then multiply $v = 0.333 \ldots$ with ten to power the number of digits that recur, that is, $10^1 = 10$.

$$\begin{aligned} \text{So;} \quad 10v &= 3.333\ldots \\ \text{Now subtract} \quad -v &= 0.333\ldots \\ \hline 9v &= 3.00 \\ v &= \frac{3}{9} \end{aligned}$$

$\Rightarrow v = 1/3$ Therefore, 0.333 ... = $1/3$ as a fraction.

Example 2

Change the following decimals to fractions (i) 0.272727 ...   (ii) 0.34111 ...

Solution

(i) In 0.272727 ... , two digits recur, therefore we multiply with the second power of ten, that is, $10^2 = 100$.

$$\text{Let } w = 0.272727 \ldots$$

$$100w = 2\,7\,.\,2\,7\,2\,7\,2\,7 \ldots$$
$$-\,w = \quad 0\,.\,2\,7\,2\,7\,2\,7 \ldots$$
$$99w = 2\,7\,.\,0\,0\,0\,0\,0\,0$$
$$w = \frac{27}{99}$$
$$\Rightarrow w = {}^3/_{11} \qquad\qquad \therefore 0.272727 \ldots = {}^3/_{11}$$

(ii) Let $x = 0.34111 \ldots$

$$10x = 3\,.\,4\,1\,1\,1\,1 \ldots$$
$$-\,x = 0\,.\,3\,4\,1\,1\,1 \ldots$$
$$9x = 3\,.\,0\,7\,0\,0\,0$$

Removing decimal point from the numerator;

$$x = \frac{3.07}{9} \times \frac{100}{100}$$
$$\Rightarrow \quad x = \frac{307}{900} \quad \therefore 0.341 = {}^{307}/_{900}$$

Or Look for powers of ten which can be multiplied by the equation $x = 0.34111 \ldots$ and easily eliminate the decimals.

If $x = 0.34111 \ldots$

$$1000x = 3\,4\,1\,.\,1\,1\,1 \ldots$$
$$-\,100x = \quad 3\,4\,.\,1\,1\,1 \ldots$$
$$900x = 3\,0\,7\,.\,0\,0\,0$$
$$\Rightarrow \quad x = \frac{307}{900} \qquad \therefore 0.341 = {}^{307}/_{900}$$

Example 3

Change the following decimal to fraction: 0.2949494 ...

Solution

Let $y = 0.2949494 \ldots$

$$100y = 2\,9\,.\,4\,9\,4\,9\,4 \ldots$$
$$-\,y = \quad 0\,.\,2\,9\,4\,9\,4 \ldots$$
$$99y = 2\,9\,.\,2$$
$$y = \frac{29.2}{99} \times \frac{10}{10}$$
$$\Rightarrow y = \frac{292}{990} = \frac{146}{495}$$
$$\therefore 0.2949494 \ldots = {}^{146}/_{495}$$

Or $1000y = 2\,9\,4\,.\,9\,4\,9\,4\,9\,4 \ldots$
$$-\,10y = \quad 2\,.\,9\,4\,9\,4\,9\,4 \ldots$$
$$990y = 2\,9\,2$$
$$y = \frac{292}{990}$$
$$\Rightarrow \quad y = \frac{146}{495}$$
$$\therefore 0.2949494 \ldots = {}^{146}/_{495}$$

Exercise 6*i*

Convert the following decimals to fractions

1	1.222 …	5	0.999 …	9	0.676777 …
2	0.333 …	6	0.8222 …	10	2.363666 …
3	0.232323 …	7	0.94111 …	11	1.892889288928 …
4	0.724724724 …	8	0.66777 …	12	0.0012232323 …

REVISION EXERCISE 6

1 Express the following as decimal numbers

   (a) $\dfrac{13}{100}$ (b) $\dfrac{7}{10}$ (c) $\dfrac{3}{5}$ (d) $\dfrac{5}{8}$

2 (a) What is the place value of 3 in the following
     (i) 0.000321   (ii) 1020.301   (iii) 3   (iv) 21.00003

  (b) Express the number 324.847019
     (i) to the nearest whole number   (ii) correct to 2 places of decimal
     (iii) to the nearest thousandths   (iv) to 1 significant figure

3 (a) Arrange the following decimals in ascending order,
     (i) 0.01, 1.00, 0.001, 10.00, 0.100   (ii) 201, 2.01, 0.201, 20.1, 2.10, 0.02

  (b) Work out: (i) 0.0042 x 100   (ii) $\dfrac{3.241 \times 1000}{10000 \times 0.3241}$

4 Work out the following:
  (a)   (i) 124.42 + 13.001   (ii) 20 + 0.349

     (iii) 100 – 39.001   (iv)   36.701 – 2.999

  (b)   (i) 13.24 x 33.61   (ii) 0.942 x 102.13

     (iii) 34.0017 ÷ 0.17   (iv)   0.4128 ÷ 0.004

  (c)   (i) $\dfrac{2.31 \times 12.101 + 1.03 - 0.302}{(0.34 \div 0.017)\ of\ 20}$ (ii) $\dfrac{(8.428 \div 0.004) - 0.39 + 6.4}{3.92 \times 100 + 8}$
     (*Express your answers in (c) above to 3 significant figures*)

5 (a) Express the following decimals as fractions
     (i) 10.34   (ii) 0.00313   (iii) 0.024
     (iv) 0.625   (v) 0.075

(b) Express the following fractions as decimals.

(i) $\dfrac{3}{5}$ (ii) $1\dfrac{1}{8}$ (iii) $\dfrac{25}{8}$ (iv) $\dfrac{2}{11}$

(v) $\dfrac{11}{12}$ (vi) $\dfrac{9}{22}$ (vii) $\dfrac{25}{3}$ (viii) $2\dfrac{2}{9}$

(c) Express the following recurring decimals to fractions:

(i) 0.8333 ...   (ii) 0.232323 ...   (iii) 0.0222 ...

(iv) 0.27111 ...   (v) 0.257257257...   (vi) 0.000323232 ...

7 PERCENTAGES

INTRODUCTION

Percentages are fractional numbers expressed with the denominator 100. If you score 82 out of 100 marks, it means that you have scored 82 percent, which is written as 82%, meaning $^{82}/_{100}$. If you scored 25 out of 40 marks in a test, your results can also be expressed as a percentage.

7.1 Expressing fractions as percentages

In this case we need to make the denominator equal to 100 without changing the meaning of the original expression.

Example 1

Change the following fractions to percentages (i) $\dfrac{2}{5}$ (ii) $\dfrac{31}{50}$ (iii) $\dfrac{17}{25}$

Solution

Make the denominator equal to 100, without changing the meaning of the original fraction.

(i) $\dfrac{2}{5} = \dfrac{2 \times 20}{5 \times 20}$

$= \dfrac{40}{100}$

$\therefore {}^{2}/_{5} = 40\%$

(ii) $\dfrac{31}{50} = \dfrac{31 \times 2}{50 \times 2}$

$= \dfrac{62}{100}$

$\therefore {}^{31}/_{50} = 62\%$

(iii) $\dfrac{17}{25} = \dfrac{17 \times 4}{25 \times 4}$

$= \dfrac{68}{100}$

$\therefore {}^{17}/_{25} = 68\%$

Example 2

Change the following fractions to percentages (i) $\dfrac{11}{40}$ (ii) $\dfrac{82}{55}$

Solution

The denominators in this case are not easily converted to 100, so we resort to just multiplying the fraction with 100 as follows;

(i) $\dfrac{11}{40} \Rightarrow \dfrac{11}{40} \times 100 = \dfrac{1100}{40} = \dfrac{55}{2}$

$= 27\tfrac{1}{2}\%$

(ii) $\dfrac{82}{55} \Rightarrow \dfrac{82}{55} \times 100 = \dfrac{8200}{55} = \dfrac{1640}{11}$

$= 149\,\text{J}/11\%$

Exercise 7a

Express the following fractions as percentages.

1 $\dfrac{1}{5}$ 2 $\dfrac{3}{8}$ 3 $\dfrac{9}{4}$ 4 $\dfrac{31}{20}$ 5 $2\,^{1}/_{3}$

6 $6\,^{1}/_{7}$ 7 $2\,^{1}/_{10}$ 8 $11\,^{3}/_{4}$ 9 $\dfrac{11}{24}$ 10 $\dfrac{55}{105}$

11 $3\,^{3}/_{5}$ 12 $12\,^{1}/_{2}$ 13 $3\,^{2}/_{100}$ 14 $\dfrac{105}{99}$ 15 $\dfrac{20}{17}$

Example 3

Change the following decimals to percentages (i) 0.75   (ii) 0.005

Solution

In this case we first change the decimals to fractions, then finally to percentages.

(i) $0.75 = \dfrac{75}{100}$ (ii) $0.005 = \dfrac{5}{1000}$ $\Rightarrow$ $\dfrac{5}{1000}$ $\dfrac{\div 10}{\div 10}$

$= 75\%$ 1000 divided by 10 makes the denominator 100, the required denominator for percentages.

$$= \dfrac{(5 \div 10)}{(1000 \div 10)} = \dfrac{0.5}{100}$$

$$= 0.5\,\%$$

Exercise 7b

Express the following as percentages

1 0.14   2 0.008   3 0.421   4 3.04   5 0.101

6 4.302   7 0.0006   8 4.4010   9 0.42   10 0.05

11 0.34   12 0.00001   13 0.1011   14 0.308   15 8.04

7.2   Expressing percentages as fractions and decimals

Example 1

Change the following percentages to fractions: (i) 54%   (ii) 250%

Solution

(i) $54\% = \dfrac{54}{100}$, canceling by 2

$54\% = \dfrac{27}{50}$

$= {}^{27}/_{50}$, as a fraction

(ii) $250\% = \dfrac{250}{100}$, canceling by 50

$250\% = \dfrac{5}{2}$

$= 2\,\frac{1}{2}$, as a fraction

Exercise 7c

Change the following percentages to fractions

1 2 %	2 40 %	3 25 %	4 80 %	5 33 %
6 12 $\frac{1}{2}$ %	7 7 $\frac{1}{2}$ %	8 66 %	9 120 %	10 75 %
11 28 %	12 $^{1}/_{8}$ %	13 4 $^{1}/_{7}$ %	14 333 %	15 77 $^{1}/_{2}$ %
16 0.04%	17 19.2 %	18 130.4 %	19 0.002 %	20 4.8 %

Example 2

Change the following percentages to decimals (i) 25 %   (ii) 234 %

(iii) 64.7 %   (iv) 72 $\frac{1}{4}$ %

Solution

Change the percentages to fractions, then to decimals.

(i) $25\% = \dfrac{25}{100} = 0.25$

$\therefore \quad 25\% = 0.25$

(ii) $234\% = \dfrac{234}{100} = 2.34$

$\therefore \quad 234\% = 2.34$

(iii) $64.7\% = \dfrac{64.7}{100} = 0.647$

$\therefore \quad 64.7\% = 0.647$

(iv) $72\frac{1}{4}\% = \dfrac{289}{4}\% = \dfrac{289}{4} \Big/ 100$

$= \dfrac{289}{400} = 0.7225$

$\therefore \quad 72\frac{1}{4}\% = 0.7225$

Exercise 4d

Change the following percentages to decimals.

1 52%	2 14%	3 44%	4 0.64%	5 12 $\frac{1}{2}$%

6 14.6%	7 8.88%	8 69%	9 $^{11}/_9$ %	10 201%
11 13.13%	12 16.24%	13 3.41%	14 11.1%	15 30.25%
16 12 ½ %	17 1 ½ %	18 34 $^1/_3$ %	19 $^{23}/_3$ %	20 $^{21}/_{100}$ %

7.3   Problems involving percentages

In a mixed school we may represent the number of boys and girls in a school population as a   percentage. For a school with a students total population of 600, with 450 boys, we can express the number of boys as a percentage of the total number of students. In this case the percentage of the boys in the school will be 75%; meaning that for every 100 students, 75 are boys.

Example 1

Two mathematics tests A and B were done. Test A was marked out of 60 and test B was marked out of 50. A student who sat for both test A and test B got 48 and 42 marks respectively.
- (a) Find his percentage marks in each of the test
- (b) In which test did he perform better?
- (c) What is his average percentage mark for both the tests?

Solution

(a) Test A , 48 out of 60. That is,   Test B, 42 out of 50. That is,

$\frac{48}{60}$ x 100, as a percentage.   That is, $\frac{42}{50}$ x 100, as a percentage.

= 80 %   = 84%

(b) Therefore, he performed better in test B.
   Note that if you merely took his results as 48 and 42 you would be misled that he did test A better than B. So it is important to compare using a similar scale, in this case percentage.

(c) Average = $\frac{80\% + 84\%}{2}$ = $\frac{164\%}{2}$ =   82%

So, his average percentage mark for both papers was 82 %.

Exercise 7e

Express each of the following as a percentage.

1 60 out of 80	2 33 out of 72	3 63 out of 90

4 11 out of 13   5   25 out of 40   6   260 out of 320

7 560 out of 720   8   0.2 out of 0.5

9   $\frac{1}{3}$ out of 4   10 $\frac{1}{4}$ out of $\frac{4}{5}$

11 In a certain school, Bob scored 32 out of 40 marks in a mid-term examination and he scored 42 out of 60 marks in the end of term examination.
   (a) Find his percentage marks in the,
      (i) mid-term examination (ii) end of term examination.
   (b) In which examination did he perform better?
   (c) What was his average percentage mark?

12 In a certain game a player obtained 3 out of 5 points in the first round, 7 out of 15 in the second round and 20 out of 25 in the third round.
   (a) Find his percentage score for each of the 3 rounds.
   (b) In which round did he perform,   (i) best?   (ii) worst?
   (c) If to win the game the player must get an average percentage score of greater than 50%, did this player win the game?

13 In a box there are marked examination papers for S.1 students. In which the students obtained the following marks;

40 out of 82	33 out of 40
32 out of 60	16 out of 100
80 out of 125	71 out of 150

   (a) What was the percentage score obtained by;
      (i) the best student?   (ii) the worst student?
   (b) Find the average percentage score for these students.

14   In a certain class, 4 students scored 50 out of 72 marks each in a test. 20 students scored 33 out of 50 marks each in another test and in the last test 18 students scored 142 out of 200 marks each.
   (a) Which group had the highest percentage score?
   (b) Find the average percentage score for the 3 groups.

15   In a basket 15 out of 80 tomatoes are rotten, 18 out of 60 mangoes are

rotten, 30 out of 125 oranges are rotten and 6 out of 100 passion fruits are rotten.

(a) Find the percentage rotten for each type of fruit.

(b) What percentage remains fresh for each fruit type?

(c) What percentage of fruits altogether are rotten?

REVISION EXERCISE 7

1 Express the following as percentages.

(i) $\dfrac{2}{7}$ (ii) $\dfrac{53}{40}$ (iii) $\dfrac{23}{36}$ (iv) $\dfrac{33}{7}$

(v) 0.0008   (vi) 0.415   (vii) 0.02022   (viii) 6.108

2 Change the following percentages to;

(a) Fractions (i) $\dfrac{3}{7}$ %   (ii) 34%   (iii) 0.68%   (iv) 72 %   (v) $2\,^3/_8$ %

(b) Decimals (i) 2 %   (ii) $^{13}/_8$ %   (iii) 2.406%   (iv) $12\,^1/_3$ % (v) $3\,^3/_7$ %

3 Express the marks obtained in various tests done by a certain student as percentage marks.

    (a)   32 out of 60   (b)   13 out of 15

    (c)   82 out of 120   (d)   99 out of 200

4 (a) Out of the 423 students who sat for a mathematics test only 243 passed the paper. What percentage of the students passed?

  (b) In mathematics tests I and II, Ivan scored 50 out of 110 and 33 out of 54 respectively;

    (i)   Find his percentage mark in each of the two papers.

    (ii) In which test did she perform better?

    (iii) Find his average percentage score.

    (iv) Express his average percentage score as a mark scored out of 40.

5 To assess a student for an end of term performance in Mofields Secondary School, the teachers consider an average of 4 tests in the term to constitute 20% of the end of term exams, 2 best done class exercises to constitute 10%

of the end of term exams and the final term examination paper to constitute the remaining percentage. If a student has the following marks:

1.	Best done exercise I		3 out of 5
2.	Best done exercise II		12 out of 20
3.	Test I		30 out of 70
4.	Test II		88 out of 100
5.	Test III		92 out of 120
6.	Test IV		60 out of 80
7.	Final term examination paper		45 out of 200

(a) Find the final percentage mark, which should be put on this studentJs report form.

(b) If the pass mark was 35% did this student pass?

(c) Do you think this student would have passed if he dodged all the exercises and tests and only final examination paper results were considered?

6   In a class of 142 students **38** $2/71$% are below 15 years, **56** $24/71$% are from 15 to 19 years and the rest are from 20 to 24 years.

(a) Find the number of students in each age bracket.

(b) If the minimum age of voting is 15 years old, how many students will not be allowed to vote?

(c) Find the percentage of voters expressed as a fraction of the total number in the class.

8 CARTESIAN COORDINATES IN 2-DIMENSIONS

INTRODUCTION

We can describe the location of a point in a given place using the co ordinate system. The system works as positioning frame in which certain points or features can be located. The framework has to be labelled with numbers from a particular known point of reference.

8.1 Location of points in a plane

Considering the arrangement of a classroom, we can locate the positions of some desks using the co ordinate system?

CLASSROOM

	Column 1	Column 2	Column 3	Column 4
Row 7	Teacher	Chalkboard		Wardrobe
Row 6				
Row 5	Ngobere	Sseruwu		
Row 4	Kamya	Nanziri	Kintu	Kato
Row 3	Mariam	Nakityo	Namutebi	Kalanzi
Row 2	Mwanje	Mukke	Wahasa	Male
Row 1	Mukasa	Musoke	Nsangi	Nayiga

Fig. 8.1

The arrangement of desks in the classroom from the left to right forms columns, and in the class we have 4 columns.

The arrangement of desks from the back to the front forms rows, and in the class we have 7 rows.

We can use these rows and columns to describe the location of pupils in the class. We shall represent our description as (column number, row number).

Example 1

Find the positions of the following pupils in class.

(i) Nakityo   (ii) Muke   (iii) Nsangi
(iv) Kalanza   (v) Naiga

Solution

(i) Nakityo is in the second column and third row.
   So we have, (2, 3) as her location in class.

(ii) Mukke is located at (2, 2)   (iv) Kalanzi is located at (4, 3)

(iii) Nsangi is located at (3, 1)   (v) Naiga is located at (4, 1)

Activity 8.1

(a) Give all the locations of the remaining students
(b) What is the teachersJ location when he/she is writing on the chalkboard?
(c) If the class monitor is standing in between the third and fourth columns and in between the fourth and fifth rows.
   (i) Locate his/her position in the class using a star in the figure showing the classroom arrangement.
   (ii) Write down the co ordinates representing his location.

Exercise 8a

1	**Row 5**				A				
	Row 4			B	C	D			
	Row 3		E	F	G	H	I		
	Row 2	J	K	L	M	N	O	P	
	Row 1	O	R	S	T	U	V	W	X	Y
		1	2	3	4	5	6	7	8	9

Columns

(a) Given the figure above, write down the co ordinates for the locations of the letters A – Y, in terms of columns and rows, for example, location A is A(5, 5).

(b) If we are to have the letter **Z** in our diagram what would be its position written in co ordinate form.

R₅	T			N		S
R₄		X			Z	
R₃				L		
R₂			P			
R₁	R				Q	
	C₁	C₂	C₃	C₄	C₅	C₆

(2)

(a) Give the coordinates for each of the letters in figure above.

(b) Locate the following letters on the figure above.

(i)   A(6, 1)   (ii)   B(4, 4)

(iii) C(1, 4)   (iv) D(6, 3)

(v) E(3, 2)   (vi) F(2, 1)

(vii) G(2, 3)   (viii) H(5.5, 1.5)

Do you think we can use this system to describe locations of houses in a certain town? Definitely we may, but it may not seem convenient since the buildings may not have a proper pattern, therefore it is better to resort to the grid system commonly known as grid reference for description of features on maps.

8.2   The Grid system

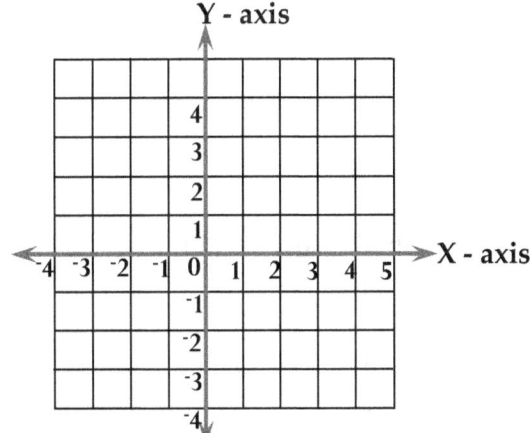

The grid system involves dividing the plane area into equal small squares, also called units.

The grid starts with two perpendicular number lines ;

Fig. 8.2

- The Horizontal axis, called the x-axis, and the
- The Vertical axis called the y-axis.

These axes are generally called coordinate axes, and they intersect at a point called the origin which is also the zero-point for both axes.

8.2.1 Labelling axes

Conventionally;

- On the x-axis, right of the origin, x-units are positive and left of the origin x-units are negative.
- On the y-axis above the origin, y-units are positive and below the origin y-units are negative.

Study the grid carefully to understand this concept of labelling the axes. We can use quadrants in a circle to further explain this concept.

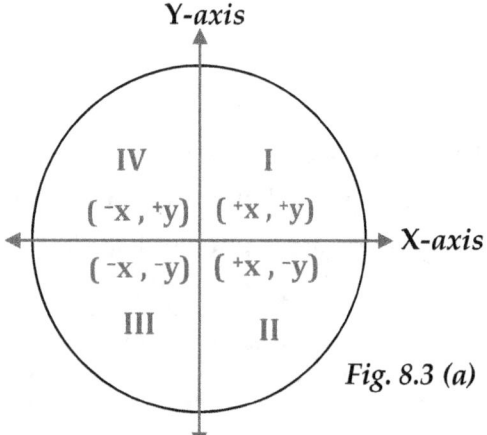

Fig. 8.3 (a)

Quadrant	Abscissa	Ordinate
I	+	+
II	+	–
III	–	–
IV	–	+

Fig. 8.3 (b)

Remember that: **X** – co ordinates are called abscissa, and the
                    Y – co ordinates are called ordinates.

8.2.2 Ordered pairs

To obtain the x-coordinate of a point, we obtain how far the point is to the right (positive) or to the left (negative) of the vertical (y-axis).

To obtain the y-coordinate of a point, we obtain how far the point is above (positive) or below (negative) the horizontal (x-axis).

We can obtain the co-ordinates of letters A – I. Remember that to write the coordinates, the x-value comes first, then the y-value comes next, and then enclose the pair separated by a comma in parentheses, for example, (2, ⁻1).

This means that the x-value is 2 and the y-value is ⁻1. For the co-ordinates of the letter indicated on the grid we have;

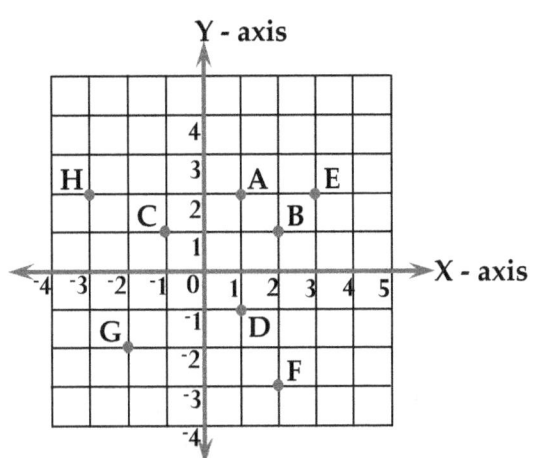

A(1, 2)   B(2, 1)   C(-1, 1)
D(1, -1)   E(3, 2)   F(2, -3)
G(-2, -2)   H(-3, 2)   I(-2, 0)

(a)   What is the difference between the coordinates of points A and B?

(b)   What is the relationship between points E and H?

Fig. 8.4

Remember that the co-ordinates for the origin are O(0, 0).

Exercise 8b

1 Given the letters on the grid find their co ordinates.

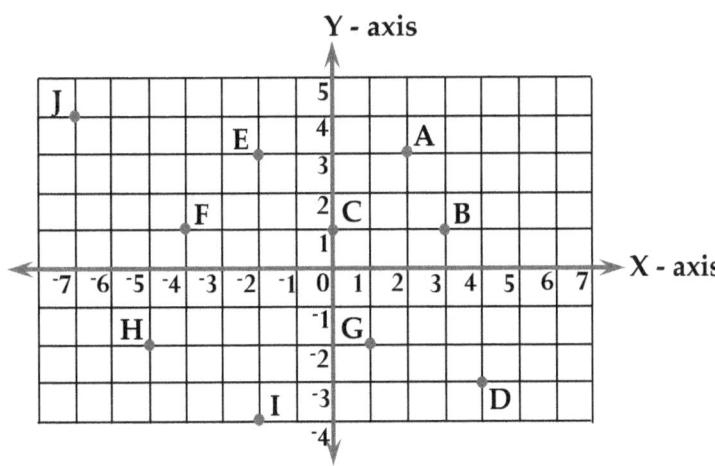

2 On the same grid using the mark "*" plot the following points.

    (i)   K(3, 2)   (ii)   L(1, 5)   (iii) M(0, 3)   (iv) N(4, 2)

    (v) P(1, -3)   (vi)   Q(3, -4)   (vii) R(0, -3)   (viii) S(-1, 2)

    (ix) T(-4, -4)   (x)   U(0, -4)

8.2.3 Scales on grids

A grid is made up of squares, and each square is taken to represent any units, say, mm or cm.

Therefore, our scales here will be in the form: 1 square represents x units; where x is any natural number on the grid. For convenience, we normally take scales like;

1 square rep. 1 unit;	1 square rep. 2 units
1 square rep. 5 units	1 square rep. 10 units and so on.

Example 1

Make a grid with a scale:
(i) 1 sq. rep. 2 units, up to atleast 8 units in all directions from zero.
(ii) 1 sq rep. 5 units, up to atleast 20 units in all directions from zero.

Solution

(i) **Scale:** 1sq. rep. 2 units on both axes

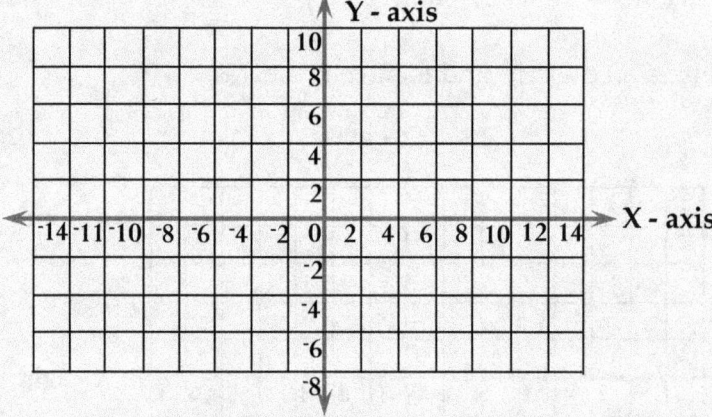

(ii) **Scale:** 1sq. rep. 5 units on both axes

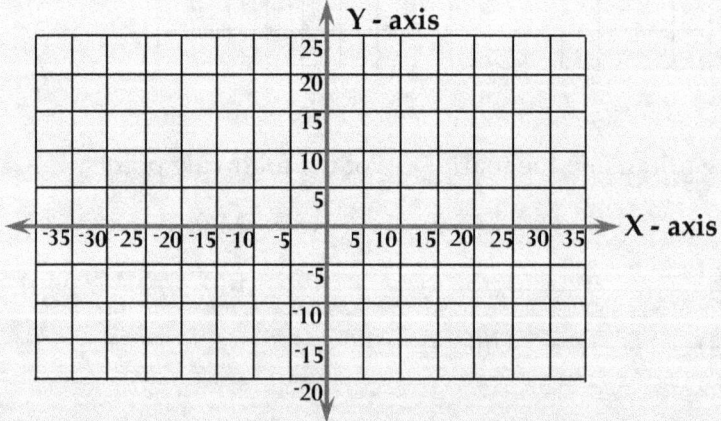

Example 1

Make a grid with scales;

1sq. rep. 1 unit on the x-axis, and

1sq. rep. 2 units on the y-axis, up to 6 units on either sides of the origin for both axes.

Solution

Scales :   1sq. rep. 1 unit on the x-axis, and

1sq. rep. 2 units on the y-axis.

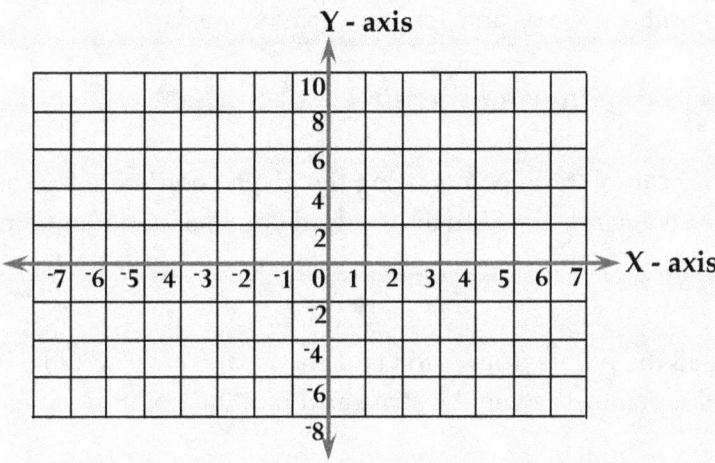

Note:

1   For standard graph papers 1 square is 1 cm by 1cm, so while using these graph papers 1 sq. rep 1 unit will be 1cm rep 1 unit.

2   For values of x to range from ⁻8 to 10 it can be written, as ⁻8 ≤ x ≤ 10. Likewise for y we have a range ⁻7 ≤ y ≤ 7; which means that the values of y are such that ⁻7 is less than or equal to y and y is less than or equal to ⁺7.

Exercise 8c

Make grids with the following scales and the given ranges of *x* and *y*-value.

1   1sq. rep 4 units; for ⁻30 ≤ x ≤ 45 and ⁻20 ≤ y ≤ 16

2   1sq. rep 5 units; for ⁻20 ≤ y ≤ 35 and ⁻20 ≤ y ≤ 35

3   1 sq. rep. 15 units; for ⁻100 ≤ x ≤ 30 and 0 ≤ y ≤ 75

---

4 1 sq. rep. 0.5 units; for $-4 \leq x \leq 2$ and $-10 \leq y \leq 10$

On a standard graph paper take approximately the middle to be the origin and mark the graph paper with scales.

5   1 cm rep. 4 units
6   1 cm rep. 10 units.
7   1cm rep. 0.5 units on x axis and 1cm rep. 2 cm on y-axis.
8   1cm rep. 8 units
9   2cm rep. 0.5 units (change this scale to 1: x first)
10 1cm rep. 5 units on x-axis and 1cm rep. 4cm on y-axis.

8.3   Further coordinate plotting (x, y) in the Cartesian plane

Given a grid we can plot the points using the given coordinate for each point. In this case you will be required to draw a grid your self and plot points on it.

Exercise 8d

Look through all the points from A to J and choose the most suitable scale that will cover all the points, then on the same grid plot the points;

1 A(3, 2)	2 B(2, 3)	3 C(-3, 2)	4 D(-2, 3)
5 E(-2, -3)	6 F(-3, -2)	7 G(2, -3)	8 H(3, -2)
9 I(5, -1)	10 J(7, 5)		

Look through all the points from A to P below and choose the most suitable scale that will cover all the points, then on the same grid plot the points;

11 A(-3, 7)	12 B(-1, 1)	13 C(3, -2)	14 D(-4, 6)
15 E(-2, 0)	16 F(8, -4)	17 G(-3, -3)	18 H(0, 1)
19 I(-2, 5)	20 J(-5, -1)	21 K(-2, 3)	22 L(-6, -2)
23 M(0, 0)	24 N(-1, -2)	25 P(-2, -3)	

8.3.1 Locating midpoints, thirds, fourths, fifths, etc.

We have been dealing with coordinates which are whole numbers and now we are going to have a look at the fractional coordinates. In this case we are going to use a standard graph paper to study this.

On a standard graph paper, each large square is 1 cm (either way– width or length). It is then subdivided into five smaller units.

If each large square is equivalent to: 1 cm

Therefore, each small square is: $(1 \div 5) = 0.2$ cm.

Example 1

Plot the following points on a grid with a scale 1sq. rep. 1 unit.

(i) A(2.5, 3)   (ii) B(-3½, 2)   (iii) C(3.2, 1.8)

(iv) D(5.4, 3.1)   (v) E(-1¼, 2 ⅕)

Solution

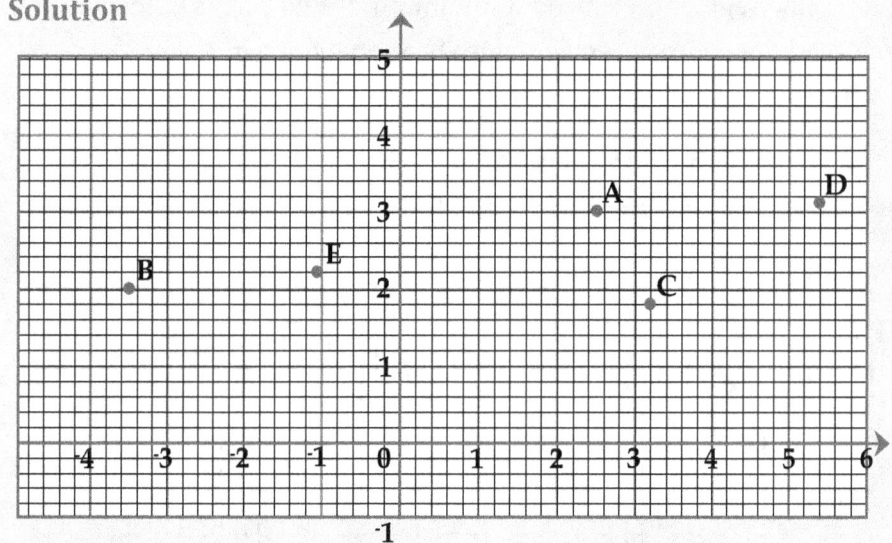

Points to note:

(i) To plot $^1/_2$; we take 5 small squares divided by 2 = 2.5 small squares.

(ii) To plot $^1/_5$; we take 5 small squares divided by 5 = 1 small square.

(iii) To plot $^1/_4$; we take 5 small squares divided by 4 = 1.25 small squares.

(The scale for this graph is: 1 cm rep. 1 unit on both axes.)

Activity 8.2

What do you think would be the procedure to plot the following?

(i)   A(2.13, 2)   (ii) B(1.25, 10)

(iii)   C(4.6, 3.8)   (iv) D(2 $^{21}/_{40}$, 3 $^3/_{10}$)

Exercise 8e

Plot the following points on the grid:

1   M(2, 4.2)   2   N($^-$3 $^3/_{10}$, 2½)   3   P(0.8, $^-$2.4)   4   Q($^-$2, $^-$1.4)

5   R($^-$5, 1 ⅛)   6   S($^-$2.2, 3.0)   7   T($^-$1.1, 3.5)   8   U(2.25, 3.4)

9   V(2 ½, 3 ⅜)   10 W(0, 0.5)

8.4   Plotting of line segments and figures on a grid

To plot a line segment, we need a minimum of two points. Other figures like geometric shapes, curves, etc., can also be plotted on a grid.

   Example 1

(a) On a grid plot the coordinates A($^-$3, $^-$1), B($^-$1, 0), C(1, 1), D(3, 2) and E(5, 3)

(b) Join the points and name the figure formed.

   Solution

(a)

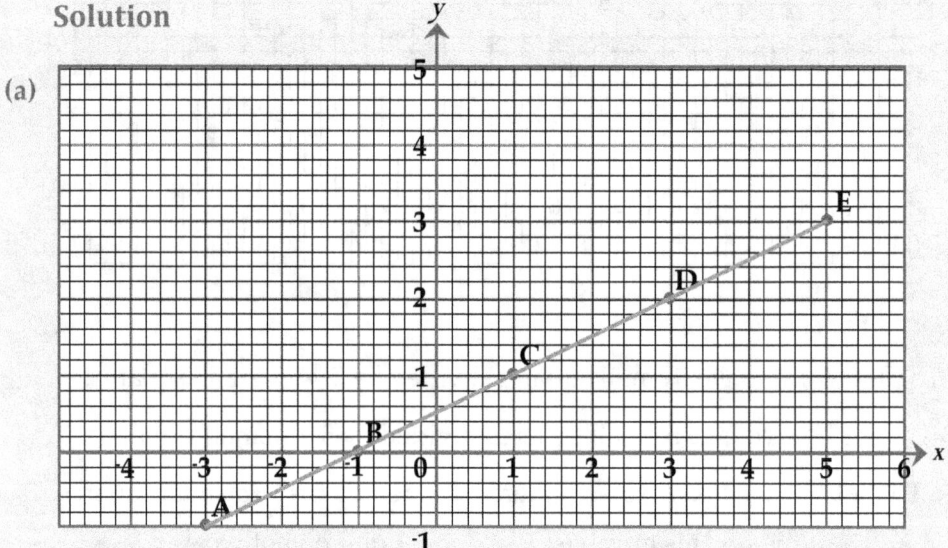

(b) The figure formed is a straight line.

   Example 2

(a) Given the points K(3, 1), L(0, 4), M(2, 6) and N(4, 4); plot them on the same grid.

(b) Join all the points K, L, M and N; and name the formed shape.

Solution

(a)

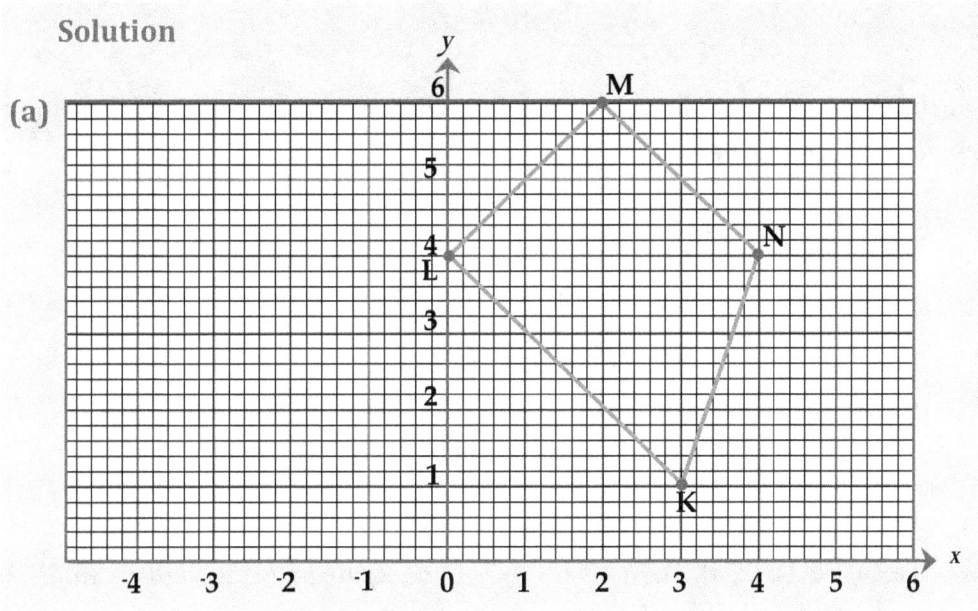

(b) The shape KLMN is a quadrilateral.

Example 2

The table gives the x and y values of a figure. On a graph draw the figure and state its name.

Solution

(a)

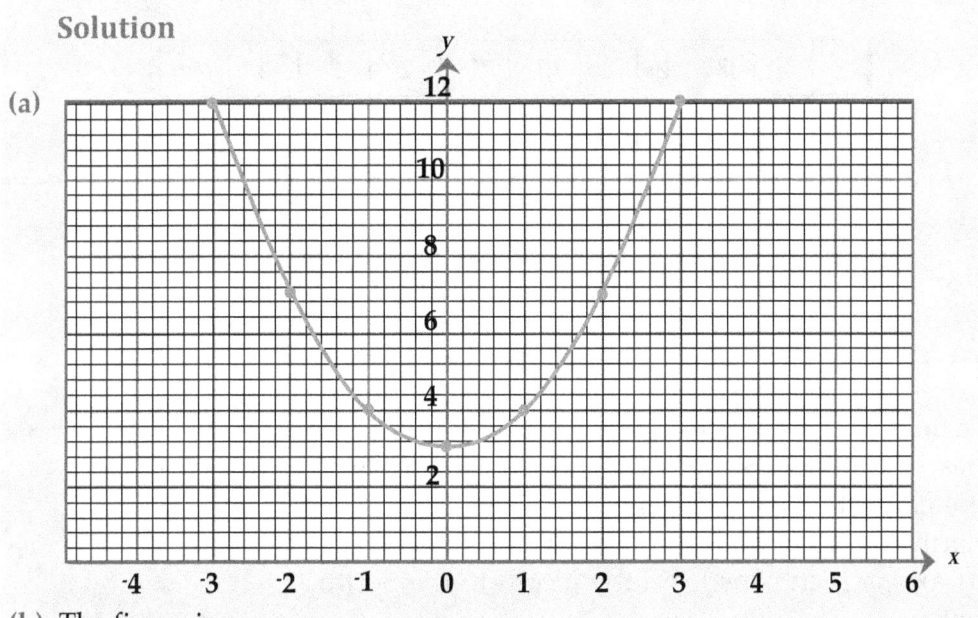

(b) The figure is a curve

Exercise 8f

On different grids, plot each of the following set of points, join them and name the figure formed.

1   (-3, 3), (-2, 3), (-1, 3), (0, 3), (1, 3), (2, 3)

2   (-2, 2), (-2, -1), (-2, 0), (-2, 1), (-2, 2)

3   (-2, -2), (-1, -1), (0, 0), (1, 1), (2, 2), (3, 3)

4   (-2, -4), (-1, -3), (0, -2), (0, 3), (1, -1), (2, 0), (3, 1)

5   (-3, -8), (-2, -5), (-1, -2), (0, 1), (1, 4), (2, 7)

6   (1, 1), (3, 3), (8, 4)   7   (1, 1), (3, 4), (4, -1)

8   (1, 0), (-1, 1), (2, 4), (4, 3)   9   (2, 2), (0, 8), (4, 10), (10, 6), (6, 2)

10 (-5, 1), (-4, 3), (-2, 3), (0, 1)

11 (-3, 14), (-2, 9), (-1, 6), (0, 5), (2, 9), (3, 14), (4, 21)

12 (-3, 31), (-2, 11), (-1, -1), (0, -5), (1, -1), (2, 11), (3, 31)

13

x	-3	-2	-1	0	1	2
y	-6	-7	-6	-3	2	9

14

x	-2	-1	0	1	2	3
y	-16	-9	-8	-7	0	19

15

x	-5	-4	-3	-2	-1	0	1	2	3	4	5
y	-26	-15	-6	1	6	9	10	9	6	1	-6

8.5 Interpretation of grid reference on maps

We can use the idea of graphs to label and locate places on maps. To know the locations of various features, the coordinate system is used. The Earth is nearly a perfect sphere and all the points on it are about the same distance from its centre. The Earth rotates about an imaginary line known as its axis, which runs from the North to the South pole.

Considering the Earth, it possesses a reference frame made of latitudes and longitudes.

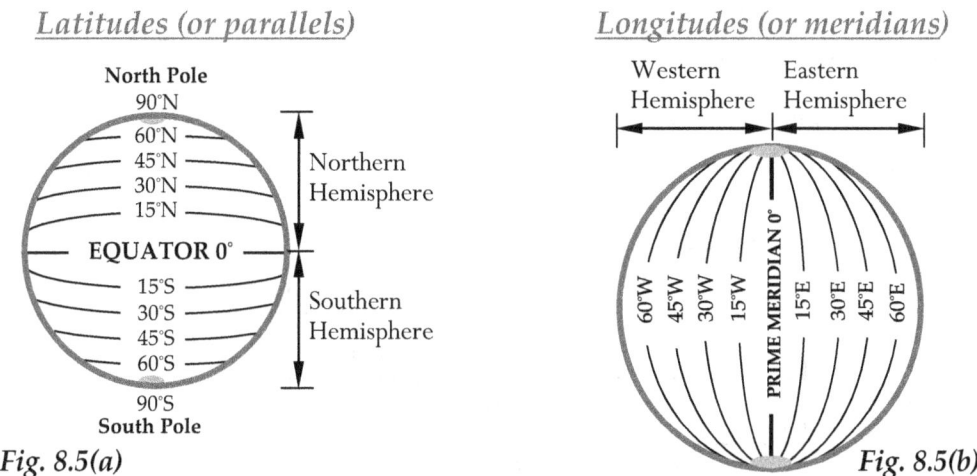

Fig. 8.5(a)   *Fig. 8.5(b)*

Every point on the equator is equidistant from the north or south pole.

If the latitudes and longitudes are combined, they form a network like that of the grid and it can also be labeled as;

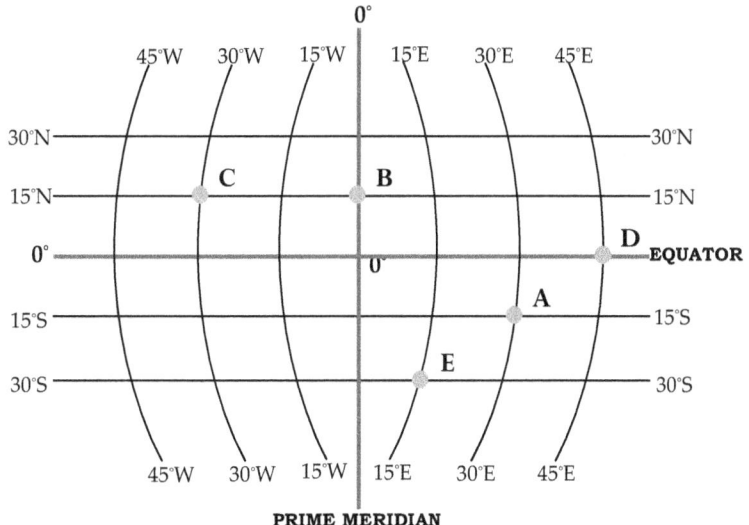

Fig. 8.6

A feature at point A is 15° South and 30° East. A feature at point B is 15° North and 0° (at the prime meridian). This kind of locating features on a map using the lines of latitude and longitude is known as grid reference. The readings can also be known as coordinates.

Activity 8.3

Write the latitudes and longitudes representing the locations of the features at points C, D and E.

Below is a map of Watson village. It has been drawn in a grid with square 1cm by 1cm. The map is drawn on scale 1:100,000.

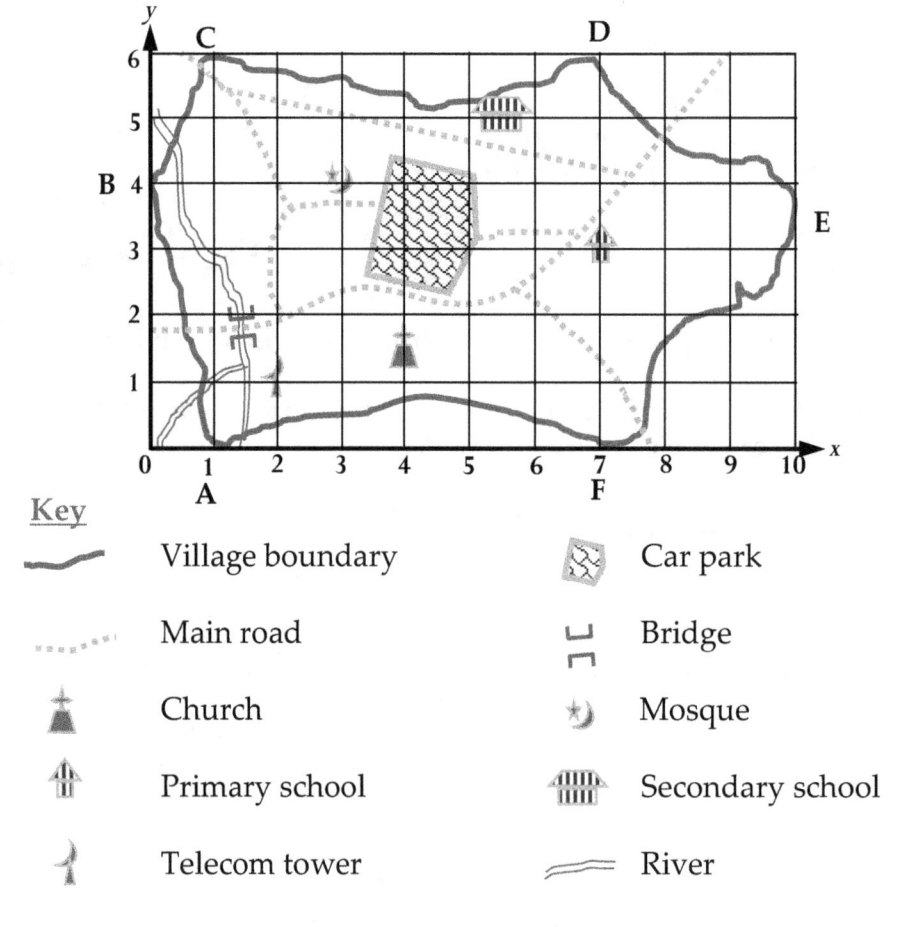

Key

—— Village boundary   Car park

····· Main road   Bridge

Church   Mosque

Primary school   Secondary school

Telecom tower   River

Exercise 8g

Give the locations of the following: 1 Secondary School.   2 Primary School.

3 Church   4 Bridge   5 Car park   6 Recreation park

7 Telecom tower   8 Primary Sch.   9 Secondary Sch.

10 Points A, B, C, D, E and F on the boundary of the map

REVISION EXERCISE 8

1 On the grid plot the following points:

   (i) A(-1, 2)   (ii) B(-3, 0)   (iii) C(3, -5)

   (iv) D(3.5, 4)   (v) E(-0.5, 0)   (vi) F(½, ¼)

   (vii) G(-¾, ½)   (viii) H(0, 0.5)

2 On the following grid, give the coordinates of the points labeled by letters.

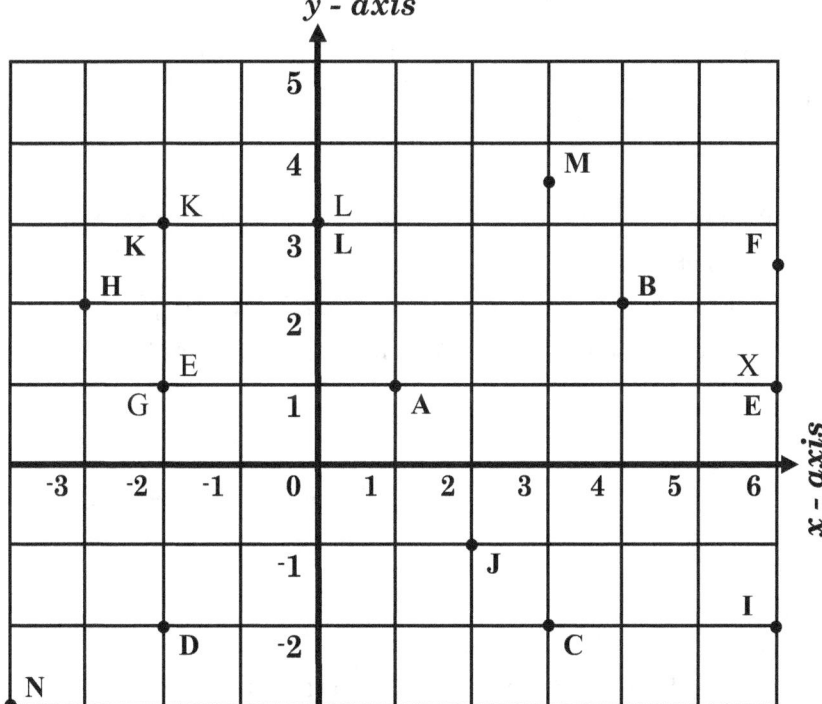

3 Given the points, A(3, 1), B(6, 2), C(-1, 0), D(-1, 3) and E (-¼, 2)

   (a) Plot them on a grid.

   (b) Find lengths of the lines: (i) AB   (ii) CD   (iii) BE

                                             (iv) AC   (v) CE

9 ALGEBRAIC SYMBOLS

INTRODUCTION

Under various instances we use letters of the English or Greek alphabet to represent values of say, distance, speed, time, angle and so on. For such instance we may have a distance x metres, a weight of y kilograms or a speed of z kilometers per hour and so on.

9.1 Symbolic representation of statements

When we se a letter to represent a number, can take a specific or various numerical values.

Example 1

(a) I am x years old and my sister is p years old. Write an expression for the sum of our ages.

(b) In a class of 40 students, 12 weigh p kgs each, 18 weigh q kgs each and 10 weigh $2r$ kgs each. Find an expression for the total weight of the class.

Solution

(a) $(x + y)$ years

(b) $\{(12 \times p) + (18 \times q) + (10 \times 2r)\}$kgs
$$= (12p + 18q + 20r)\text{kgs}$$

Example 2

At the market stall there are 32 tomatoes, 18 tomatoes weigh x grammes each, 12 tomatoes weigh y grammes each and the rest of the tomatoes weigh $2t$ grammes each. Write the expression for;

(i) Total weight of the tomatoes.   (ii) Average weight of the tomatoes.

Solution

(i)   Total weight. Remember for the rest of the tomatoes;
        we subtract $(18 + 12)$ from 32

$\{18x + 12y + [32 - (18 + 12)]2t\}$ grammes

$$= \{18x + 12y + 2t(32 - 30)\} \text{ grammes}$$
$$= \{(18x + 12y + (2t \times 2))\} \text{ grammes}$$
$$= (18x + 12y + 4t) \text{ grammes}$$
$$= 2(9x + 6y + 2t) \text{ grammes}$$

(ii) Average weight is given by: Total weight of tomatoes
Total number of tomatoes.

i.e. $\left[\dfrac{2(9x + 6y + 2t)}{32} \right]$ grammes

$= \left[\dfrac{(9x + 6y + 2t)}{16} \right]$ grammes

Example 3

A man traveled x km in 4 hours, y km in 2h hours and $3z$ kms in 3m hours. Write an expression for;

(i)   The total distance he traveled.

(ii) The total time he took traveling.

(iii) His average speed for the whole journey.

Solution

(i) Total distance traveled: $(x + y + 3z)$ km.

(ii) Total time taken: $(4 + 2h + 3m)$ hrs.

(iii) Average speed; Total distance traveled $= \left(\dfrac{(x + y + 3z)}{(4 + 2h + 3m)} \right)$ Km/hr
Total time taken

Exercise 9a

1   There are x English books, $2w$ Language books, 4 Mathematics books and $(x + 2)$ Physics books. What is the expression for the total number of the books altogether?

2 Four students weigh $2p$ kgs, xy kgs, 42 kgs and $(2p - 5)$ kgs.

  (a) Write an expression for their total weight.

  (b) Write an expression for their average weight.

  (c) Write an expression for the difference in weights $2p$ kgs and $(2p - 5)$ kgs where $2p$ kgs is the greater weight.

3 In a bookshelf with 20 books, 4 weigh $(2x + 1)$ kgs each, 10 weigh w kgs each and the rest weigh $\dfrac{(w + x)}{2}$ kgs each.

  (a) Write an expression for the total weight of the books weighing $\dfrac{(w + x)}{2}$ kgs each.

  (b) Write an expression for the total weight of all the 20 books.

  (c) Write an expression for the average weight of all the 20 books.

4 To empty the water tank we needed 7 containers of $b(x + 2)$ litres each; $(p + 2)$ containers of $(6x - 7)$ litres each; 4 containers of $(2p + x)$ litres each and $(2x - 7)$ containers of $n(x + 3)$ litres each. Write an expression for;

  (i) The total number of litres the tank can hold altogether.

  (ii) The average number of litres each container can hold.

5 In a kraal there are $(4x + 3)$ bulls weighing $(40p + q)$ kgs altogether, $(35x + 1)$ cows weighing $b(y + 2x)$ kgs altogether and np calves weighing $(33x - 2)$ kgs altogether. Write an expression for;

  (i)   Weight per bull

  (ii) Weight of $\dfrac{2x}{b}$ cows

  (iii) Total weight of animals in the kraal.

  (iv) The average weight of all the animals in the kraal.

6 In a class 14 pupils have $(2x - y)$ Bic pens each, $(4x + 1)$ pupils have 2 Nice pens each and $(n + 1)$ pupils have $(12y - x)$ Picfare pens.

  (a) Write an expression for:-

    (i) The total number of pens in the class.

(ii) The average number of pens per person in the class.

(b) If there are more Bic pens than the Nice pens by how many does the Bic pens exceed the Nice pens?

(c) If the total number of pens in the class was increased by $(2x - p)$ Bic pens, what would be the total number of Bic pens in the class?

9.2 Number properties

9.2.1 Commutative property

Addition

When adding numbers, their order of arrangement does not change the sum. If we have a sum $(23 + 67)$ and $(67 + 23)$, they will both give the same answer.

That is; $23 + 67 = 90$ and
$67 + 23 = 90$

Generally for commutative property of addition we have: $(a + b) = (b + a)$
In similar cases, we may have:

$$(4 + 3) = (3 + 4) = 7 \; ; (2 + 7) = (7 + 2) = 9; (5 + 4) = (4 + 5) = 9, \text{ etc..}$$

Multiplication

When we multiply numbers their order of arrangement does not change the product. If we have a product (20×31) and (31×20) they will both give the same answer.

That is; $20 \times 31 = 620$ and
$31 \times 20 = 620$

Generally for commutative property of multiplication: $(a \times b) = (b \times a)$

$$ab = ba$$

In similar cases, we may have;

$$(2 \times 3) = (3 \times 2) = 6; (3 \times 14) = (14 \times 3) = 42; (5 \times 3) = (3 \times 5) = 15; \text{ etc.}$$

Note the similarity between addition and subtraction for the commutative property.

9.2.2 Associative property

Addition

If we have a sum of say, (4 + 18 + 6); we can associate 4 and 18 first, then associate the result, 22, with the remaining number, 6.

On the other hand we can associate 18 and 6 first, then associate the result, 24, with the remaining number, 4.

We ca also first associate 4 and 6; that is, (4 + 6) + 18

$$So, (4 + 18) + 6 \ = \ 4 + (18 + 6) \ = \ (4 + 6) + 18$$

Remember what the parentheses mean, that is;

$$(4 + 18) + 6 \ = \ 4 + (18 + 6) \ = \ (4 + 6) + 18$$
$$22 \ + 6 = 4 + \ 24 \ = \ 10 \ + 18$$
$$28 \ = \ 28 \ = \ 28$$

Generally for the associative property of addition we have:

$$(a + b) + c \ = \ a + (b + c) = (a + c) + b$$

Multiplication

If we have a multiplication say, (2 x 3 x 4); we can associate 2 with 3 first, then associate the result, 6, with the remaining number, 4.

On the other hand we can associate 3 and 4 first, then associate the result, 12, with the remaining number, 2. Or associate 2 and 4 first, then 3

$$So, (2 \times 3) \times 4 \ = \ 2 \times (3 \times 4) \ = \ (2 \times 4) \times 3$$

Remember,
$$(2 \times 3) \times 4 \ = \ 2 \times (3 \times 4) \ = (2 \times 4) \times 3$$
$$6 \ \times 4 = 2 \times \ 12 \ = \ 8 \ \times 3$$
$$24 \ = \ 24 \ = \ 24$$

Generally for the associative property of multiplication we have;

$$(a \times b) \times c \ = \ a \times (b \times c) \ = (a \times c) \times b$$

9.2.3 Distributive property

If we have three groups of fives and three groups of fours, we can sum them up as follows:

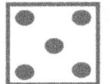 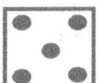   PLUS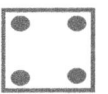

3 groups of **5** beads each   +   3 groups of **4** beads each

3 x 5   +   3 x 4

Have you realized that the 3 is common in both products, that is;

$$(3 \times 5) + (3 \times 4) = 3(5 + 4)$$

Generally for the distributive property; $(a$ **x b)** $+ (a$ **x c)** $= a$ **(b + c)**

I n this case, Multiplication is distributed over Addition

Checking: $(3 \times 5) + (3 \times 4)$ = $3(5 + 4)$

$(15) + (12)$ = $3(9)$

27 = 27

Activity 9.1

State *true* or *false*

(a) $4 + 3 + 7 + 9 = 7 + 9 + 4 + 3$

(b) $8 \times 21 \times 11 = 21 \times 8 \times 11$

(c) $8 - 6 = 6 - 8$

(d) $12 + (8 + 3) + 1 = 3 + 1 + (8 + 12)$

(e) $2 \times (11 \times 7) \times 3 = (7 \times 3) \times 11 \times 2$

(f) $8(3 + 2 + 1) = 8 \times 3 + 8 \times 2 + 8 \times 1$

9.3   Simplifying Algebraic expressions, brackets and their systems

We normally use parentheses in algebra to indicate the operation to be done first. If we have $6(3 + 9)$, it means that we need to add first, then multiply the sum by 6.

That is, $6(3 + 9) = 6(12)$

$= 72.$

Remember how we dealt with BODMAS in Chapter two.

Still if we remember BODMAS, given an expression like: $122 - 13 \times 6 + 9 , 3$, we need to use parentheses to show what we must do first and so on.

This can be done in two ways, one is right and the other is wrong.

$\underline{122 - 13} \times 6 + 9 \div 3$	$122 - 13 \times 6 \ + 9 \div 3$
$\underline{109 \times 6} + 9 \div 3$	$122 - 13 \times 6 \ + (9 \div 3)$
$\underline{654 + 9} \div 3$	$122 - (13 \times 6) + 3$
$\underline{663} \div 3$	$122 - \quad 78 \quad + 3$
$= 221 \textbf{ WRONG}$	$(122 + 3) - \quad 78$
	$125 - 78$
	$= 47 \textbf{ CORRECT}$

Given an expression $a(x + y)$ where a-value is multiplied with a bracket, we can open or remove the brackets, as follows;
$$a\,(x + y) \quad = \quad a\,(x + y)$$
$$= \quad ax + ay$$

Here to open the parentheses we have multiplied the value JaJ by all the terms in the bracket. This process is called *expanding*.

For example, given $2(4 + 5)$ we shall have;
$$2(4 + 5) = 2 \times 4 + 2 \times 5$$
$$= \quad 8 \ + 10$$
$$= \quad 18$$

Remember the distributive property of multiplication over addition.

Example 1

Expand the following: (i) $b(x + y + z)$ 　　(ii) $3(5 + y)$

Solution

(i) $b(x + y + z) \quad = \quad b.x + b.y + b.z$
$$= \quad bx + by + bz$$

(ii) $3(5 + y) \quad = 3.5 + 3.y$
$$= 15 + 3y$$

Example 2

Expand the following: (i) $(p + q - 3)b$ 　　(ii) $^-2(p - 2y)$

Solution

(i) $(p + q - 3) = b.p + b.q - b.3$
$$= bp + bq - 3b$$

(ii) $^-2(p - 2y) \quad = \quad ^-2.p + ^-2.\,^-2y$
$$= \quad ^-2p + 4y$$
$$= \quad 4y - 2p$$

We may be given a bracket multiplied by another bracket, for example $(a + b)\,(x + y)$.

In this case we multiply the first term, a in the first bracket with both the terms x and y in the second bracket.

Then multiply the second term, b in the first bracket with both terms x and y in second bracket.

$$(a + b)(x + y) = a(x + y) + b(x + y)$$
$$= ax + ay + bx + by$$

Example 3

Expand the following: (i) $(a + p)(m + n)$ (ii) $(p - q)(r + t)$

(iii) $(m - 3)(2p - 1)$

Solution

(i) $(a + p)(m + n) = a(m + n) + p(m + n)$ (ii) $(p - q)(r + t) = p(r + t) + {}^-q(r + t)$

$$= am + an + pm + pn$$
$$= p(r + t) - q(r + t)$$
$$= pr + pt - qr - qt$$

(iii) $(m - 3)(2p - 1) = m(2p - 1) + {}^-3(2p - 1)$

$$= 2p - m - 6p + 3$$

Example 4

Expand the following; (i) $-2(m + 2)(3 + r)$ (ii) $(a + b)(p - q)(3 - m)$

Solution

(i) $-2(m + 2)(3 + r) = -2[m(3 + r) + 2(3 + r)]$

$$= -2[3m + mr + 6 + 2r]$$
$$= {}^-6m - 2mr - 12 - 4r$$

(ii) $(a + b)(p - q)(3 - m) = [\,a(p - q) + b(p - q)\,](3 - m)$

$$= (ap - aq + bp - bq)(3 - m)$$
$$= (3 - m)(ap - aq + bp - bq)$$
$$= 3(ap - aq + bp - bq) - m(ap - aq + bp - bq)$$
$$= 3ap - 3aq + 3bp - 3bq - amp + amq - bmp + bmq.$$

Exercise 9b

Expand the following;

1	$2x(x - 3)$		2	$3(p - q + 3)$
3	$2x(3 - m)$		4	$3p(b - yx)$
5	$^-3p(2 - p)$		6	$8n(x - y) - 3x(1 - p)$
7	$pq(8 + 6m) - (6 + 2p)$		8	$3pm(3m - 2y) + 38(2p - 1)$
9	$2pn(n - 1) + 16(pn - 2)$		10	$(2 - x)(x - yz)$
11	$(3y - 2p)(4q - r)$		12	$3p(x + 2)(2p + y)$
13	$(8y - 3p)(34 - 2p)(6 - 7n)$		14	$3m(pn - qr)(qr - p)$
15	$(42 - 3q)(b - pn)(5 - 6p)z$			

9.3.1 Constants and Variables

If you know the total weight of 20 eggs of similar size, then you can find out the weight of each egg. In this case you may represent the unknown weight with a letter then workout to find its value. If the weight of 20 eggs is 1600grammes altogether, we may let the weight of one egg be represented by a letter, y for easy working. The value of y is *constant*, that is, the value of $y = 80$gms for all instances.

If a tank losses water continuously, then its weight will keep changing. This therefore means that the weight of the tank is not fixed. Its weight can not be a constant value. It is better to represent it with a letter, say, w kgs. In this case w is called a *variable*, since it has various answers.

Given an algebraic expression, $2x - 5$; 2 and 5 are constants, then x is a variable.

9.3.2   Terms and Coefficients

Values separated by the addition sign "+" or the subtraction sign "−", are called *terms*. For example $2x$, xq, yzp, ^-4xy, etc. are each a single term.

An expression with only one term is called a *monomial*. For example $2x$, xq, yzp, ^-4xy, etc. are monomials.

An expression with two terms is called a *binomial*. For example, $(m + n)$, $(x - 2p)$, $(x^2 - w)$, etc. are binomials.

An expression with three terms is called a *trinomial*. For example, $(2x^2 - y + 2px)$, $(pq^2 - yp + mn^2)$, $(xy^2 + zp - 2w)$, etc. are trinomials.

Monomials or a sum of monomials, (i.e. binomials, trinomials etc.) are called *polynomials*.

For the terms $2x$, px and $3qx$, if x is a variable while 2, p and $3q$ are constants, then we have 2, p and $3q$ as *coefficients* of x in the terms $2x$, px and $3qx$ respectively. If a term has no number (constant) part as a coefficient then its coefficient is 1, for example, the coefficient of the term y is 1.

Like terms

Terms with similar variables are said to be *like terms*, e.g. the pairs of terms; $2x$ and $39x$, $3y^2$ and $6y^2$, $2xy^2z$ and $19xy^2z$, etc. are each like terms.

In the terms $2x$ and $39x$, the similar variable is x; in the terms $3y^2$ and $6y^2$; the similar variable is y^2 and in the terms $2xy^2z$ and $19xy^2z$, the similar variable is xy^2z.

Collecting like terms

For like terms say, $32z^2$ and $2z^2$, we can combine them to obtain $34z^2$, that is, to collect like terms we add the coefficients of the like terms.

Example 1

Collect like terms in the following: $2p - 3qx + p + 9qx + 21p$

Solution

First arrange the polynomials such that all the like terms are put together.

$$2p - 3qx + p + 9qx + 21p \quad = \quad 2p + 21p + p - 3qx + 9qx$$
$$(2 + 21 + 1)p + (^-3q + 9q)x$$
$$24p + (6q)x$$

Example 2

Collect like terms in the following: (i) $2x^2 + xy - 3xy + 3x + 4x^2$

(ii) $3y^2 - y + 2xy + 3y + 6y^2 + yx$

Solution

(i) $2x^2 + xy - 3xy + 3x + 4x^2$

$2x^2 + 4x^2 + xy - 3xy + 3x$

$6x^2 - 2xy + 3x$

Note that x^2 and x are not like terms, though they have a common factor x.

(ii) $3y^2 + 6y^2 - y + 3y + 2xy + yx$

$3y^2 + 6y^2 - y + 3y + 2xy + yx$

$(3 + 6)y^2 + (\text{-}1 + 3)y + (2 + 1)xy$

$9y^2 + 2y + 3xy$

Note that xy and yx are like terms since multiplication of numbers is commutative. However, it is preferred to express the terms in alphabetical order, that is, we prefer xy to yx; xyz to yzx; abc to acb or cba or bca, etc.

While dealing with polynomials you need to always remember that;

(i) A variable with no number or letter in front of it, its coefficient is 1, e.g. $x = 1x$, so the coefficient of x is 1. In the term px, the coefficient of x is p where x is the variable and p is the constant.

(ii) Only terms with similar variables (like terms) can be added or subtracted.

(iii) To add terms we add their coefficients and maintain their variables, which variables in all instances must be the same.

Exercise 9c

Collect like terms in each of the following:

1	$(a + b) + 3b - 1$	2	$2ab - 3b + ab - b$
3	$xyz - pz + 2xyz$	4	$1 - 42p + 3y + 32p - y$
5	$3y^2 - 3y + 4y^2 - xy$	6	$xp - 2px + 3zp - yx$
7	$8pq - 4lm - n + 3pq - n$	8	$5xy + xyz - 2yx - bx$
9	$6pq - qp + mn - xyz$	10	$32pz - qp - 3ml - hzp.$

9.3.3 Word phrases and algebraic expressions

We can use various phrases to mean a particular algebraic expression. Study the following word phrases and their corresponding algebraic expressions.

Concept	Word phrase	Algebraic expression
—	• A number	n
Addition	• A number plus 1 • A number added to 1 • 1 more than a number • A number increased by 1 • A number exceeded by 1	$n + 1$
Subtraction	• A number minus 5 • A number less than 5. • 5 less than a number. • Take away 5 from a number • Subtract 5 from a number	$n - 5$
Multiplication	• 2 times a number. • A number multiplied by 2 • The product of 2 and the number • Twice the number • Double the number	$2n$
Division	• A number divided by 8 • An nth, eg $1/8$ is an eighth. • The quotient of the number and 8 • 8 into the number • A number over 8 • Reciprocal of a number • 1 over n	$n/6$ or $1/6\,n$ $1/n$
Exponential	• A number squared • A number power 2 • The second power of the number • A number times itself • A number multiplied by itself • A number raised to the power 2	n^2
Exponential	• A number power $1/x$ • X th root of the number • A number power negative x.	$n^{1/x}$ n^{-x} $x\sqrt{n}$

Follow all the representatives of word phrases and note how they work.

Remember that the following signs will act as instructions used in the word phrases;

Equal sign, = ;

Less than sign, < ;   *Greater than sign, > ;*

Greater than or equal sign, ≥ ; and   *Less than or equal sign, ≤ .*

Exercise 9d

Write expressions for the following word phrases:

1   *"the reciprocal of a number minus 7"*

2   *"the reciprocal of twice the number"*

3   *" number reduced by 5 reciprocated"*

4   *" the eighth root of a number"*

Write word phrases for the following expressions:

5 $2x - 1$ 6 $\dfrac{x^2}{2}$ 7 $n^{\frac{1}{2}} - \dfrac{1}{n}$ 8 $\dfrac{3(n-2)J}{n}$

Write expressions for the following phrases;

9 *"Half the square of the number was reduced by twice its reciprocal"*

10 *"A number was raised to the third power, increased by 4, the result halved and reduced by half the number squared".*

9.4   Algebraic fractions

We have already looked at fractions in chapter five, however; here we shall look at fractions with either a denominator, numerator or both as or with a letter(s), for example;

$$\frac{2}{x} \, , \quad \frac{y}{5} \, , \quad \frac{x}{y} \, , \quad \frac{2x}{p} \, , \quad \frac{1}{2y} \, , \quad \frac{x-2}{p} \, , \quad \frac{3p}{7y-2} \, , \quad \text{etc.}$$

Example 1

Simplify :-   (i) $\dfrac{12y^3}{2y^4}$ (ii) $\dfrac{30x - x^2}{2x}$

Solution

(i) $\dfrac{12y^3}{2y^4} = \dfrac{12 \times y \times y \times y}{2 \times y \times y \times y \times y} = \dfrac{6}{y}$

(ii) $\dfrac{30x - x^2}{2x}$ Factorize out x in the numerator since it is a factor of both $30x$ and x^2

$\dfrac{30x - x^2}{2x} = \dfrac{x(30 - x)}{2x}$ Now cancel x to obtain;

$= \dfrac{[30 - x]}{2}$

Example 2

Simplify :- $\dfrac{(x - 2)J + (x - 2)}{(x - 2)J}$

Solution

$\dfrac{(x - 2)J + (x - 2)}{(x - 2)J}$

$\dfrac{(x - 2)[(x - 2) + 1]}{(x - 2)J}$ Factorise out $(x - 2)$ in the numerator since it is common for both terms in the numerator.

$\dfrac{(x - 2)\,[(x - 2) + 1]}{(x - 2)(x - 2)(x - 2)}$ Now cancel the factor common in both the numerator and the denominator, that is, $(x - 2)$.

$\dfrac{[x - 2 + 1]}{(x - 2)J}$ $= \dfrac{(x - 1)}{(x - 2)J}$

Exercise 9e

Simplify the following expressions

1	$2xy - 3xy$	2	$\dfrac{3x(4y - p)}{12xy - 3xp}$	3	$\dfrac{8p - zp}{3p}$
4	$\dfrac{6xy + 3x}{2y + 1}$	5	$\dfrac{3pqr - 9p}{2l - n}$	6	$\dfrac{4mn - 8ml}{qr - 3}$

7 $$\frac{(x-3p)J + (x-3p)J}{(x-3p)J}$$

Extract expressions from the following word phrases and simplify them:

8 (a) A number raised to the fourth power, reduced by twice itself and the result divided by double its square.

(b) Twice the cube of the number, reduced by eight times its square and the result multiplied by four times the square of the number.

9.4.1 Addition and Subtraction of Algebraic expressions

Example 1

Add: (i) $\frac{2}{x} + \frac{1}{x}$ (ii) $\frac{y}{x} + \frac{1}{x}$

Solution

In this case we take the same procedure as in chapter 2, only that here we deal with numbers mixed with letters.

(i) $\frac{2}{x} + \frac{1}{x}$ Find the L.C.M of x and x which is obviously x. Remember addition of fractions with same denominator.

$$\frac{2}{x} + \frac{1}{x} = \frac{2+1}{x} = \frac{3}{x}$$

(ii) $$\frac{y}{x} + \frac{1}{x} = \frac{y+1}{x} = \frac{(y+1)}{x}$$

Example 2

Add: (i) $\frac{7}{x} + \frac{y}{2}$ (ii) $\frac{3}{y^2} + \frac{1}{y}$

Solution

(i) $\frac{7}{x} + \frac{y}{2}$ L.C.M of x and 2 is $2x$

$$\frac{7}{x} + \frac{y}{2} = \frac{(2 \times 7) + xy}{2x} = \frac{14 + xy}{2x}$$

(ii) $\frac{3}{y^2} + \frac{1}{y}$ L.C.M of y^2 and y is y^2

$$\frac{3}{y^2} + \frac{1}{y} = \frac{3+y}{y^2} = \frac{(3+y)}{y^2}$$

Activity 9.2

We can work out: $\dfrac{2}{x} + \dfrac{1}{x}$ as $\dfrac{2}{x} \bowtie \dfrac{1}{x}$ to obtain $\dfrac{2x+x}{x^2} = \dfrac{x(2+1)}{x^2} = \dfrac{3}{x}$

Where " $\longleftrightarrow$ " signifies multiplication. Now work out (ii) in example 1, and (i) and (ii) in example 2 using the same method.

Exercise 9f

Simplify the following expressions

1. $\dfrac{px}{y} + \dfrac{xq}{l}$

2. $\dfrac{3p}{mn} + \dfrac{px}{2n}$

3. $\dfrac{2}{x} + \dfrac{3p}{x} + \dfrac{1}{x^2}$

4. $\dfrac{3yz}{pq} + \dfrac{2}{3p}$

5. $\dfrac{xy}{pm} + \dfrac{1}{y} + \dfrac{2p}{3m}$

6. $\dfrac{2zp}{xy} + \dfrac{3pl}{my}$

7. $\dfrac{xpq}{2} + \dfrac{x\,y}{5}$

8. $\dfrac{2xy}{pr} + \dfrac{4y}{7} + \dfrac{3}{x}$

9. $\dfrac{3y}{pz} + \dfrac{3p}{4} + \dfrac{1}{6}$

10. $\dfrac{p}{2} + \dfrac{3q}{4} + \dfrac{6}{5x}$

Example 3

Subtract: (i) $\dfrac{3}{y} - \dfrac{1}{x}$ (ii) $\dfrac{12}{x} - \dfrac{13}{2x}$ (iii) $\dfrac{2}{z} - \dfrac{1}{x}$

Solution

(i) $\dfrac{3}{y} - \dfrac{1}{x}$ L.C.M of x and y is xy. $\dfrac{3}{y} - \dfrac{1}{x} = \dfrac{3x - y}{xy}$

(ii) $\dfrac{12}{x} - \dfrac{13}{2x}$ L.C.M of x and $2x$ is $2x$. $\dfrac{12}{x} - \dfrac{13}{2x} = \dfrac{24 - 13}{2x} = \dfrac{11}{2x}$

(iii) $\dfrac{2}{z} - \dfrac{1}{x}$ L.C.M of z and x is xz. $\dfrac{2}{z} - \dfrac{1}{x} = \dfrac{2x - z}{xz}$

Activity 9.3

We can work out: $\dfrac{3}{y} - \dfrac{1}{x}$ as: $\dfrac{3}{y} \underset{\longleftrightarrow}{\times} \dfrac{1}{x}$ to obtain $\dfrac{3x - y}{xy}$

Where "$\longleftrightarrow$" means multiplication. Work out (i) and (ii) in example II above using this method.

Exercise 9g

Simplify the following expressions;

1. $\dfrac{16}{pq} - \dfrac{2x}{pm}$

2. $\dfrac{34p}{29q} - \dfrac{3r}{50}$

3. $\dfrac{2x}{y} - \dfrac{3p}{5y}$

4. $\dfrac{4zq}{5pl} - \dfrac{2l}{y}$

5. $\dfrac{34y}{21p} - \dfrac{7}{3px}$

6. $\dfrac{8lm}{xyz} - \dfrac{13xy}{pqz}$

7. $\dfrac{1}{p} - \dfrac{1}{pqr}$

8. $\dfrac{3zp}{2q} - \dfrac{1}{5qr}$

9.4.2 Multiplication and division of algebraic expressions

Example 1

Simplify: (i) $\dfrac{2x}{y} \times \dfrac{p}{2l}$ (ii) $\dfrac{3}{z} \times \dfrac{2x}{y}$

Solution

(i) $\dfrac{2x}{y} \times \dfrac{p}{2l} = \dfrac{2xp}{2ly} = \dfrac{xp}{ly}$ (ii) $\dfrac{3}{z} \times \dfrac{2x}{y} = \dfrac{6x}{yz}$

Example 2

Simplify: (i) $\dfrac{3p}{l} \div \dfrac{3x}{2}$ (ii) $\dfrac{6l}{7} \div \dfrac{2x}{h}$

Solution

(i) $\dfrac{3p}{l} \div \dfrac{3x}{2} = \dfrac{3p}{l} \times \dfrac{2}{3x} = \dfrac{2p}{lx}$ (ii) $\dfrac{6l}{7} \div \dfrac{2x}{h} = \dfrac{6l}{7} \times \dfrac{h}{2x} = \dfrac{6lh}{14x} = \dfrac{3lh}{7x}$

Exercise 9h

Simplify the following

1. $\dfrac{3xyh}{3pmq} \times \dfrac{3(p-q)}{2yl}$

2. $\dfrac{(2-x)p}{mn} \times \dfrac{2xmy}{2m-xm}$

3. $\dfrac{3p-4mn}{3 \times lp} \times 3(xp - 4l \times pq)$

4. $\dfrac{24pl}{7\,(8y-x)} \div \dfrac{py}{(x-8y)}$

5. $\dfrac{4lmn}{(2xl - 6pql)b} \div \dfrac{mn\,(p+q)}{2bl\,(93pq - x)}$

9.5   Evaluation of algebraic expressions

All the terms which are variables may be substituted with numbers and then the value of the expression may be obtained by applying various operations.

Example 1

Given that $a = 2$, $x = 6$, $y = 1$ and $z = {}^{-}2$, evaluate the following expressions;

(i) $\dfrac{2a-y}{x}$

(ii) $\dfrac{3\,(x-4y)}{z}$

(iii) axy^2z

Solution

(i) $\dfrac{2a-y}{x}$

$\dfrac{2(2)-1}{6}$

$\dfrac{4-1}{6} = \dfrac{3}{6}$

$= \dfrac{1}{2}$

(ii) $\dfrac{3(x-4y)}{z}$

$\dfrac{3(6 - 4 \times 1)}{{}^-2}$

$\dfrac{3(2)}{{}^-2} = \dfrac{6}{{}^-2}$

$= {}^-3$

(iii) axy^2z

$2 \times 6 \times (1)^2 \times {}^-2$

$12 \times 1 \times {}^-2$

$12 \times {}^-2$

$= {}^-24$

Activity 9.4

If $a = {}^-2$, $b = 3$, $c = 5$, $d = 9$

Simplify the expression $\dfrac{4d}{b} + \dfrac{2a}{cd}$ and hence evaluate.

(First carry out the addition)

Exercise 9*i*

Given that, $a = {}^-1$, $b = 2$, $c = 0$, $l = {}^-3$, $m = 6$, $x = 5$ and $y = {}^-10$, evaluate the following expressions;

1. $\dfrac{2b - 7l}{2}$

2. $\dfrac{xy + 2ab}{3}$

3. $\dfrac{aym}{3b} + \dfrac{6mx}{5b}$

4. $\dfrac{36pzc}{2} - \dfrac{abx}{y}$

5. $\dfrac{xyb}{2a} + \dfrac{4am}{lb}$

6. $\dfrac{cbx}{my} + \dfrac{3py}{19ab}$

7. $\dfrac{xy - ab}{3m}$

8. $\dfrac{8my - ab}{2x} - \dfrac{2py}{m}$

9. $\dfrac{24ym \div 6b}{xa} - \dfrac{3b \times 6}{4bm} + abmync$

10. The number of text books in a book shelf is $(3xp + 4)$ and the number of those bought is $(4x - p)$.

(a) (i) Write an expression for the sum of the old and new books altogether.

   (ii) Write an expression for the difference between the books in the shelf and those bought, provided that the old books are more than the new books?

(b) If $x = 20$ and $p = 2$, find   (i)   the sum of the books altogether.

                                         (ii)   the difference between the new and the old books.

REVISION EXERCISE 9

1 (a) In an interview Sarah scored $(x + 2)$ marks in mathematics, $(y - 1)$ marks in English, $(3x - y)$ marks in humanities and $(70 - x)$ marks in general science. Write an expression to represent her total marks.

(b) If 4 packs weigh $(p + q)$kgs, $(2p - l)$kgs, $3l$ kgs and $(50 - q)$kgs. Write an expression for;   (i)   total weight of the 4 packs.
                                        (ii)   average weight of the 4 packs.

2. In a community of 48 parents 18 possess $(4x - 3)$ children each,

12 possess $(x - y)$ children each, 6 posses $(2y + x)$ children each and the rest posses x children each.

(a) Write an expression for the;
   (i) total number of children in the community
   (ii) average number of children in the community per parent.

(b) If $6(x - y)$ children died in the village, write the expression for the average number of children in the community per parent.

3 Expand the following expressions;

(a) (i) $2a (x + y - z)$ (ii) $b(2x - p)$
    (iii) $3bc (c - p)$ (iv) $3cd (x + y) - 6p (2p - q)$

(b) (i) $(a + b) (x + y)$ (ii) $2b (p - q) (x - p)$
    (iii) $(2w - l) (m - 3p)$ (iv) $3b (z - y) (a + b) (x - p)$

4 (a) Collect the like terms: (i) $b(x - y) - x (y + b) - by$
                                   (ii) $xp - 6p + 2xp - 3p - q$

(b) Write an expression for each of the following word phrases.
   (i) *"Five times the reciprocal of a number minus twice the number"*
   (ii) *"A number plus its reciprocal and square the results"*

(c) Write word phrases for the expressions   (i) $\left(n^2 - \dfrac{n}{3}\right)^3$

                                               (ii) $\left(\dfrac{1}{n} + 1\right)^{10}$

5 Simplify: (i) $\dfrac{3pq (6 - p)}{(p - 6)q}$ (ii) $\dfrac{xyz - 3yz}{zy (b + c)}$

                (iii) $\dfrac{(2p - y)^2 + (2p - y)}{(y - 2p)}$ (iv) $\dfrac{4pq - 8pn}{2n - q}$

6 Work out: (a) (i) $\dfrac{1}{x} + \dfrac{1}{y^2}$ (ii) $\dfrac{xy}{pq} + \dfrac{p}{qy}$

                 (iii) $\dfrac{2b}{py} - \dfrac{2x}{yz}$ (iv) $\dfrac{3bc}{16} - \dfrac{12}{4cd}$

(b) (i) $2xyp \times \dfrac{1}{4pq}$ 　　　(ii) $\dfrac{24pq}{xy^2} \times \dfrac{yzw}{12q}$

 　　(iii) $\dfrac{3pt^2}{y^3} \div \dfrac{tlq}{py^2}$ 　　　(iv) $\dfrac{2pqr}{x^2y} \div \dfrac{2r^2}{x^{1/2}y}$

7　Given that $a = {}^-1$, $b = 2$ and $c = 2a$, evaluate;

(i) $2a - c$ 　　　　(ii) $\dfrac{abc - 2b}{a}$

(iii) $\dfrac{bc - 2a}{ab}$ 　　　(iv) $\dfrac{4b - ac}{2abc}$

8　In a school there are $(2x - y)$ boys and $(5y - 3)$ girls.

(a) (i) Write an expression for the total number of pupils in the school.

　(ii) If there are more boys in the school, write an expression for the number by which the number of boys exceeds the number of girls.

(b) If $x = 250$ and $y = 50$, find;

　(i) the number of boys in the school.

　(ii) the number of girls in the school.

　(iii) the total number of pupils in the school.

GEOMETRIC CONSTRUCTION

INTRODUCTION

We can use a number of geometrical set instruments to draw lines, angles and plane shapes of various properties.

You are already familiar with the use of a ruler, a protractor and a pair of compasses. In addition to the mentioned instruments, we also have set squares and a pair of dividers.

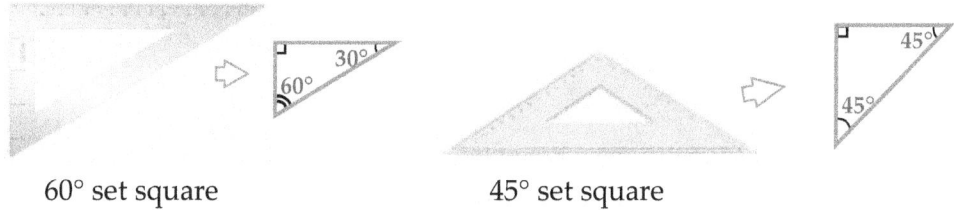

60° set square   45° set square

Dividers are used under various activities to transfer drawn distances. For example, if you have drawn a line segment 4cm long and you want to redraw it, you need not measure again. A divider can be of help.

Set squares can solve a number of drawing problems as we are about to see in this chapter.

10.1 Drawing using set squares and a ruler

A 60° or 30° set square is that longer one with 3 angles measuring; 30° – 60° – 90°. A 45° is measures 45° – 45° – 90°.

10.1.1 Drawing a perpendicular to a line

Drawing a perpendicular at any point (The 90° angle of a set square is used).

Given line AB;

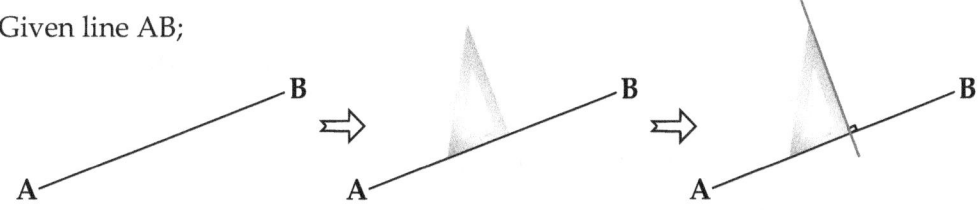

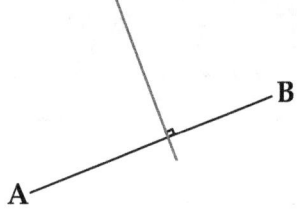

Place 1 edge of the set square against the line AB as shown. Then use the other adjacent edge to draw a perpendicular to line AB.

Perpendicular on line AB

Given a point, P on line

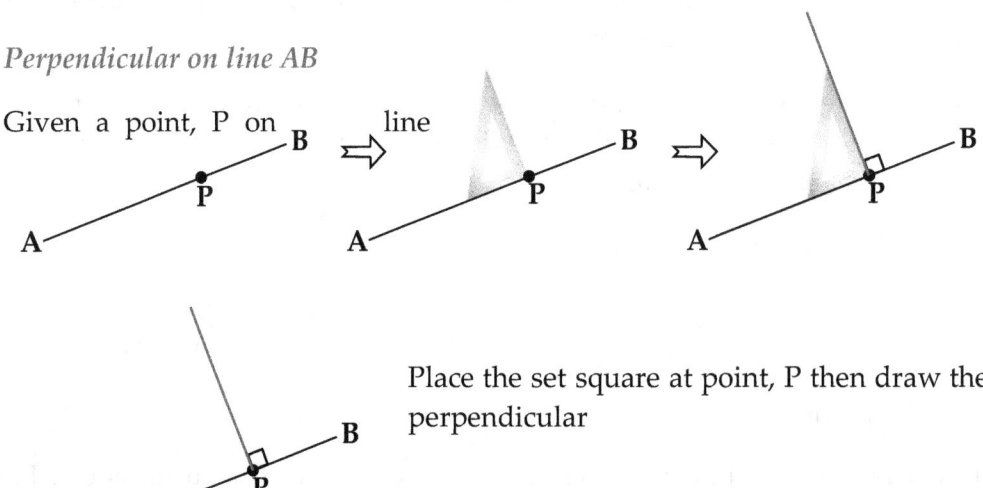

Place the set square at point, P then draw the perpendicular

10.1.2 Dropping a perpendicular to a line from a point

Given a point, P above line CD;

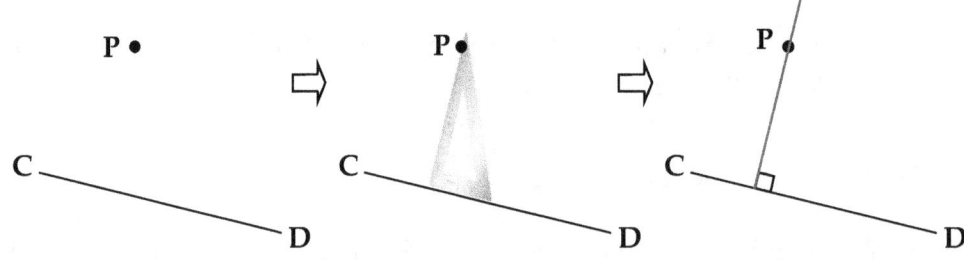

Place the set square such that the edge adjacent to that on the line passes through the point, P.

Perpendicular dropped from point, P to the CD.

Activity 10.1

Given the line PQ shown below, use a set square to;

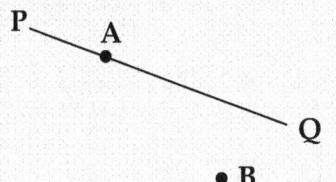

(i) draw a perpendicular at point A.

(ii) draw a perpendicular from point B to the line.

10.1.3 Drawing parallel lines

Given a line LM, we can draw another line AB parallel to line LM at any distance. In this case we need a ruler and a set square.

Draw a line LM;

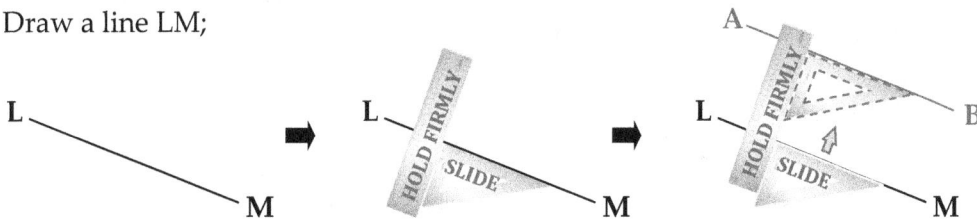

Place one edge of the set square along the line LM to be transferred. Place the ruler on the set square as shown Above.

Slide the set square along the fixed ruler to any convenient position and draw line AB along the set square.

Line LM is parallel to line AB.

Given a line, we can draw another line parallel to it and 4cm away.

Draw line XY

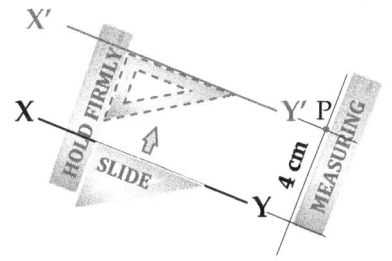

Place the set square as shown on line XY.

Place a ruler on the other edge as shown.

Using another ruler measure off. 4 cm and away from line XY. Mark a point, P = 4 cm.

Now slide the set square along the ruler up to point, P.

Exercise 10a

Given the lines below, draw lines parallel to each of them.

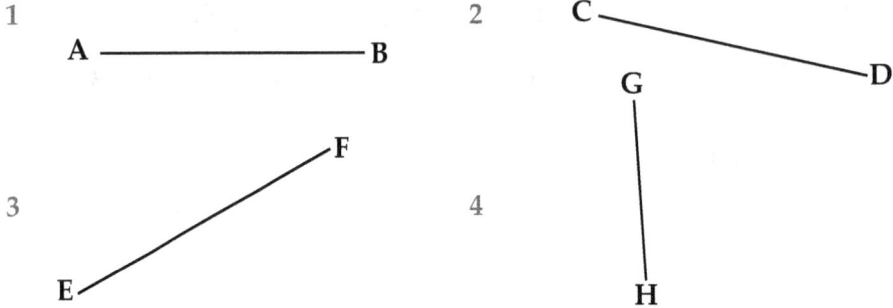

1
A ─────── B

2
C
G
D

3
F
E

4
G
H

Given the lines below, draw lines parallel to them at the point, P in each case (use a tracing paper to trace the lines).

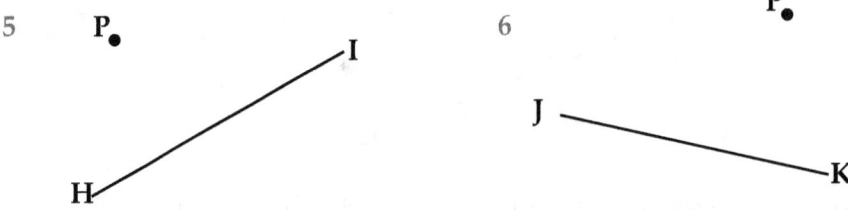

5   P
    I
    H

6   P
    J
    K

7   Draw two parallel lines AB and PQ, 6.5cm apart.

8   Given a line VW, 4.5cm long, draw it and another line XY parallel to it but 5cm away.

10.1.4 Drawing special angles

We can use set squares to draw angles. Remember that the angles of set squares are 30°, 45°, 60° and 90°. However, combined angles can also be drawn, say 75° (30° + 45°), 150° (60° + 90°) and so on.

Example 1

Given a line AB below use a set square to construct an angle of 30° at point X.

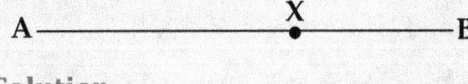

A ──────── X ──── B

Solution

Always remember the technique of sliding the set square along a ruler.

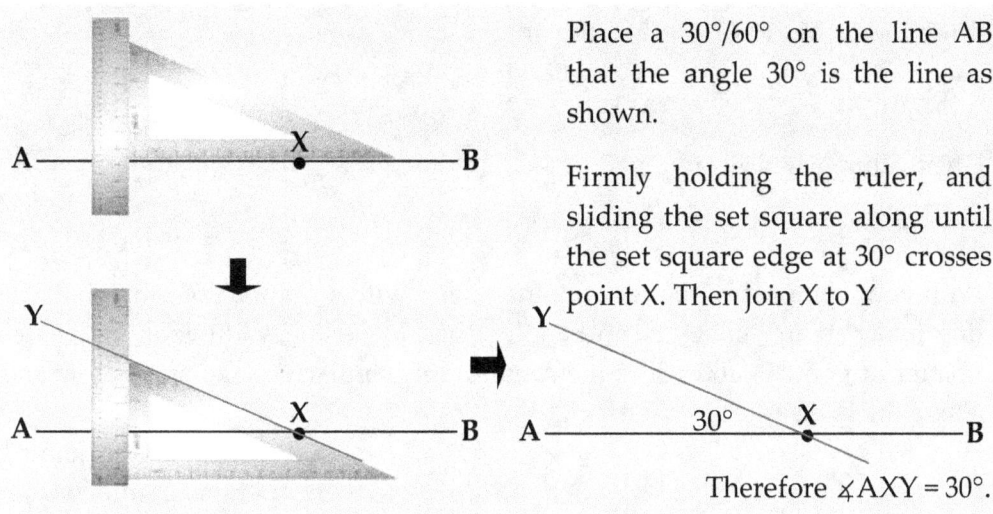

Place a 30°/60° on the line AB that the angle 30° is the line as shown.

Firmly holding the ruler, and sliding the set square along until the set square edge at 30° crosses point X. Then join X to Y

Therefore ∡AXY = 30°.

To draw an angle of 150°, use the two set squares in a combination as shown below;

Combine the 60° - corner of the 60° set square with the 90° - corner of the 45° set square as shown below. This makes 60° + 90° equalling to 150°.

Firmly hold the 2 set squares in position and align them onto the line with the two corners meeting at point, P on the line

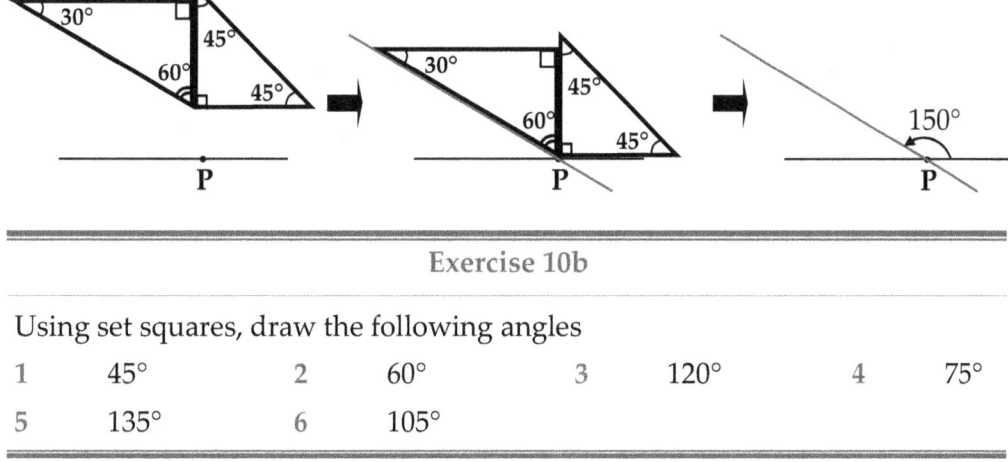

Exercise 10b

Using set squares, draw the following angles

1	45°	2	60°	3	120°	4	75°
5	135°	6	105°				

10.2 Construction of figures

Other figures can be drawn using only a pair of compasses and ruler. This is known as construction.

10.2.1 Line bisector (or Mediator)

Example 1

Given a line AB bisect it.

Solution

With your compasses pointer at point A, and with a radius more than half the line, make an arc above line and another below the line. With your compasses pointer at point B, and using the same radius similarly make arcs below and above the line.

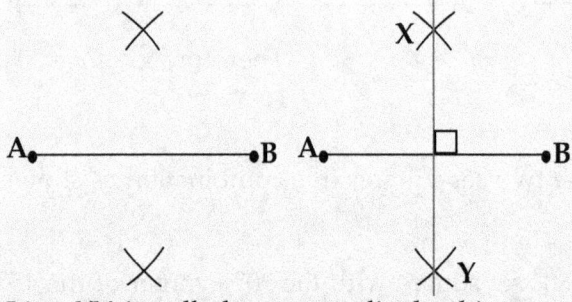

Now draw a line joining the points of intersection of arcs above and below the line AB, at X and Y respectively.

Line XY is called a perpendicular bisector for the line AB. Line XY is also called a mediator.

Activity 10.2

(a) Draw a line PQ = 70mm long and bisect it with a mediator MN crossing PQ at point X and such that a line MX = 4cm and line NX = 4.5cm.

(b) Measure and give the lengths of (i) PM   (ii) NQ

10.2.2 Parallel lines

Example 1

Given a line PQ construct another line parallel to it.

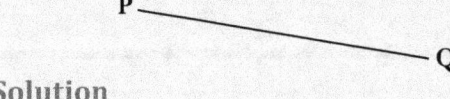

Solution

Using a pair of compasses, stretch a convenient distance between pencil and pointer. Place the compasses pointer at any three points on the line.

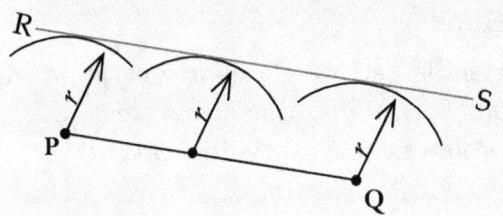

Then draw a line to just touch all the arcs line RS is parallel to line PQ.

Activity 10.3

(a) Given a line LM = 6cm, construct another line OP, 4 cm away and parallel to line LM. *(Hint: Make the radius r = 4cm)*

(b) Given the figure ABCD below, draw another figure AJBJCJDJ; 5 cm away and parallel to it.

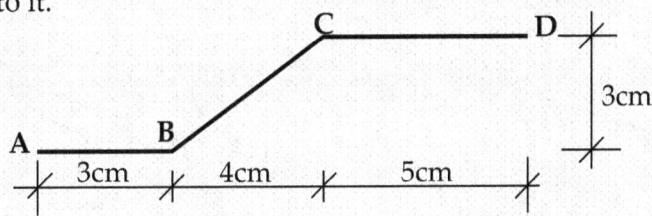

10.2.3 Angle bisectors

Example 1

Given the angle ABC below, bisect it.

Solution

With a pair of compasses pointer at point B, stretch to a convenient radius and make arcs *a* and *b*. At the point where the arcs at a and b cut lines AB and BC, make arcs meeting at *y* with a convenient radius.

Draw an angle bisector from point B to the point *y*, the intersection of arcs. ∡ABy = ∡CBy and 2∡ABy = 2∡CBy = ∡ABC

Activity 10.4

Using a protractor, draw the following angles in the figure and bisect each of the angles ACB, ACD and CDE.

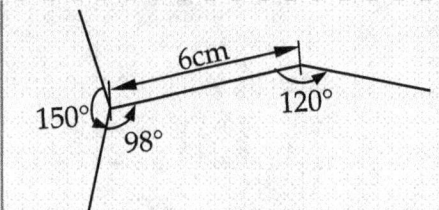

Hint:

Handle each angle as a separate angle and after you have bisected, use a protractor to check your work.

10.2.4 Dropping a perpendicular

A perpendicular can be dropped from any point to a given line.

Drop a perpendicular from point P to line AB.

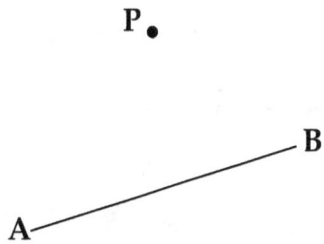

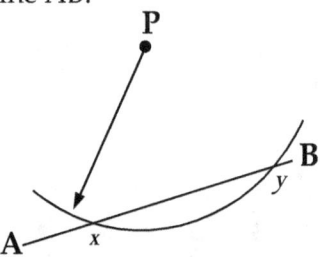

With any convenient radius (beyond the line) from point P, make an arc cutting the line AB twice as shown at points x and y. If the arc does not cut the line twice, prolong it with a dotted line until the arc crosses it.

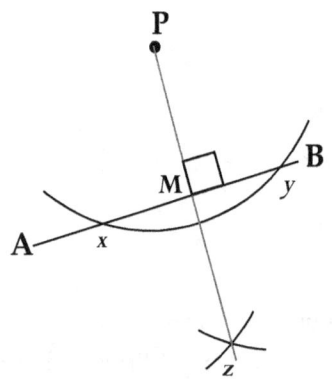

From points x and y, make arcs of the same radius to meet at z as shown.

Draw a line from point P to the point z crossing line AB at M as shown in the figure.

Therefore, $\sphericalangle AMP = \sphericalangle BMP = 90°$

Activity 10.5

Given the line QR below, drop a perpendicular onto it from the point P.

P •

Q ———————————— R

Construction of special angles

These angles are those we can construct using only a ruler and pair of compasses and not a protractor.

To construct a 90° angle, we shall only bisect a line (page 172) or drop a perpendicular (page 174).

Constructing a 60° angle

On line AB, construct a 60° angle at point X.

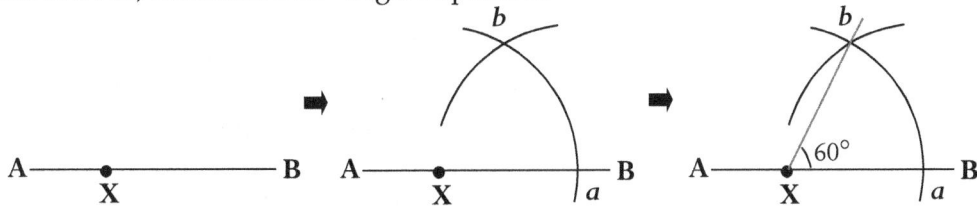

With your compasses pointer at X make an arc (of convenient radius) to cross line AB at *a*. With your compasses pointer at *a*, make an arc of the same radius to cross the first arc at *b*.

Now join point X with the point *b*. This line makes 60° with line AB.

To draw an angle of 120°, when you draw 60°, then the remaining angle at that point on a straight line is 120°.

To draw an angle of 30°, just bisect the 60° angle.

Exercise 10c

Construct the following angles

1 45°   2 22¹/₂°   3 15°   4 150°   5 112¹/₂°

10.3 Drawing triangles

10.3.1 Drawing triangles–Given all the three sides

Example 1

Draw a triangle with sides 3 cm, 6 cm and 7 cm.

Solution

Sketch

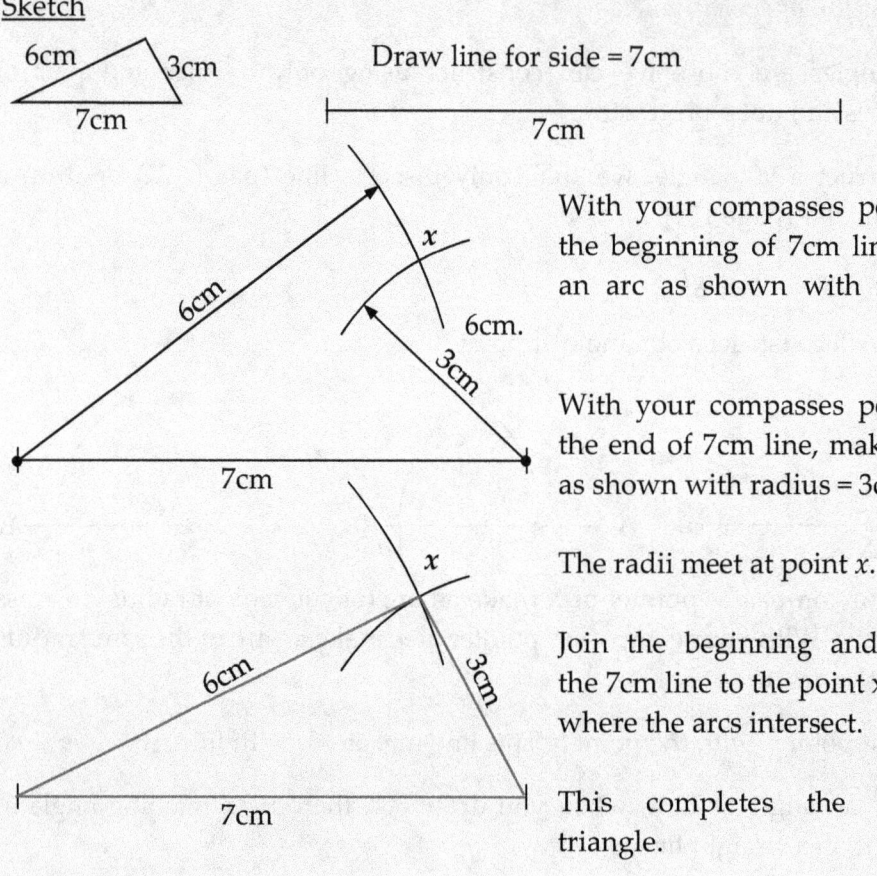

Draw line for side = 7cm

7cm

With your compasses pointer at the beginning of 7cm line, make an arc as shown with radius = 6cm.

With your compasses pointer at the end of 7cm line, make an arc as shown with radius = 3cm.

The radii meet at point *x*.

Join the beginning and end of the 7cm line to the point x, where the arcs intersect.

This completes the required triangle.

Activity 10.6

Accurately draw the triangle:

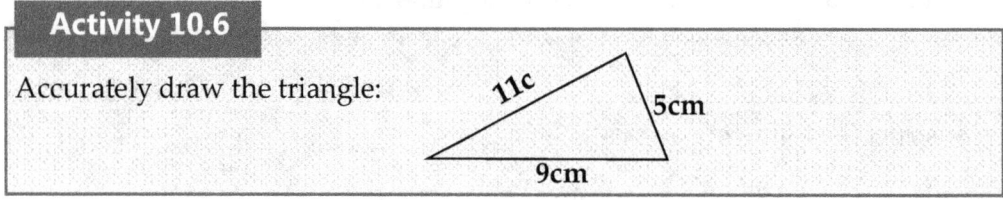

10.3.2 Drawing triangles–Given two sides and an angle between them

Given the triangle as shown below, we can redraw it accurately.

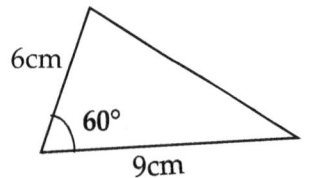

Measure and state the length of the third angle.

Draw line 9cm long

9cm

Construct the 60° angle.

The third side is: 7.8cm
(by measurement)

By calculation;

The side is: 7.94cm

6cm

7.8cm

60°

9cm

Activity 10.7

(a) Draw a triangle PQR with ∢PQR = 50°, side PQ = 7cm and line QR = 4cm.

(b) Redraw the figure below accurately;

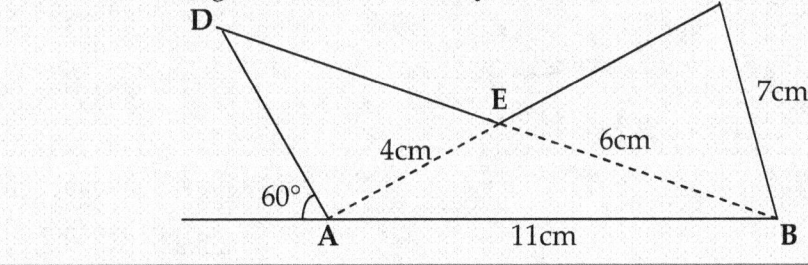

D

C

E

7cm

4cm

6cm

60°

A

11cm

B

10.3.3 Drawing triangles–Given two angles and a side between them

R

60°   45°

S   T

5 cm

Redrawing the triangle RST
accurately, we have;

We can also measure and state the length of sides RS and RT.

First draw the line ST = 5cm:

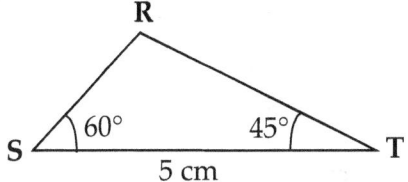

S

5cm

T

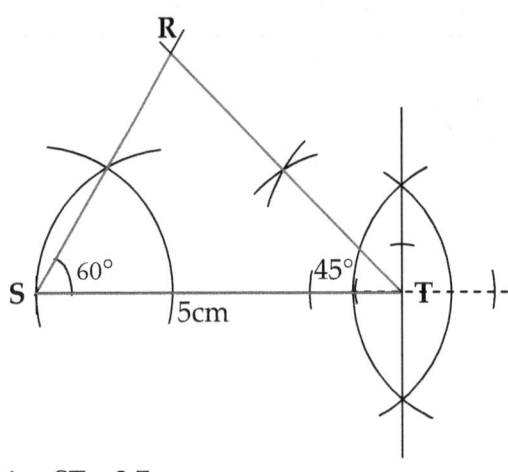

Construct a 60° angle at point S.

Construct a 90° angle at point T, and the bisect it to obtain an angle of 45°.

The line from points S and T will meet at point R.

Joint points R, S and T to obtain the triangle RST accurately drawn.

Line ST = 3.7cm   Line RT = 4.5cm

Exercise 10d

Redraw the following figures accurately, measure and give the angles and sides labeled with letters.

1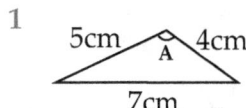

2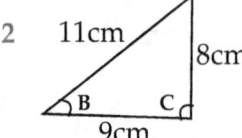

3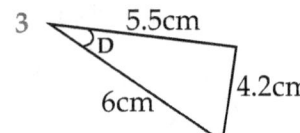

4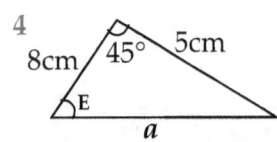

5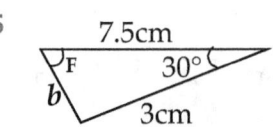

6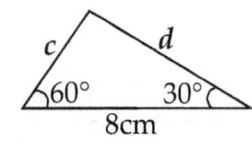

REVISION EXERCISE 10

1 Draw lines PQ = 5cm long and RS = 7cm long such that points P and R are aligned, and that they are parallel and 5cm apart.

2 Redraw the figure below accurately, and draw another figure parallel to it and 7cm away.

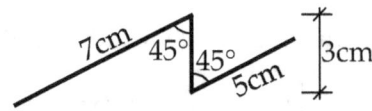

3 Using set squares, draw an angle of 105°.

4 Line AB is 9cm long. From point P, a perpendicular is dropped to this line at the 6cm mark. If the perpendicular distance from point P to the line is 5cm;

   (a) Draw the line AB and the perpendicular dropped on it from point P.

   (b) Measure and state the lengths AP and BP

Given the following figures, accurately draw them and state the unknowns represented by letters in each case.

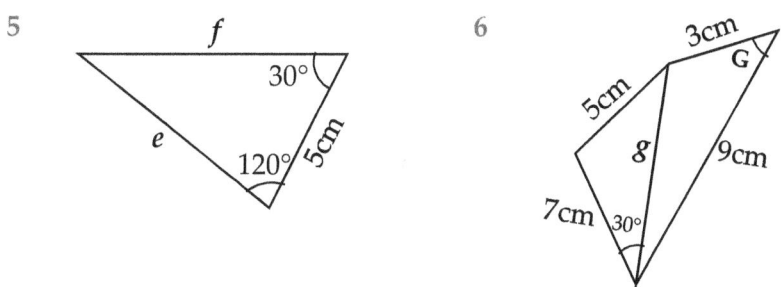

THE CIRCLE

INTRODUCTION

In our daily life, there are a lot of things which take the form of a circle, therefore there are a number of operations carried out involving circles. Such operations are basically from the concept of a circle.

A wheel, round tin of or a compact disc are examples of items derived from a circle.

When you look at this music CD, how many circles can you see?

11.1 The circle concept

A circle can be defined as a line curving around and joins up with itself in a plane, which is of equal distance from a single fixed point.

This curved and closed line is the circle and the fixed point about which it moves is its centre.

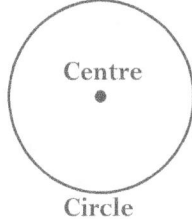

The centre of a circle is a fixed point from which the distance to any part of the closed curve forming the circle is constant.

The circumference of a circle is the distance measured around the curve which makes the circle.

11.1.1 Circle Parts: *According to lines*

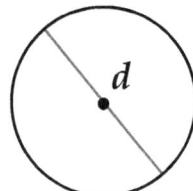
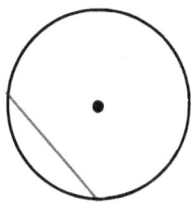

Radius of length, r   *Diametre of length, d*   *Chord*

The radius of a circle is any straight line from the centre to the curve.

The diametre of a circle is any straight line passing through the centre and touching the either ends of the curve forming the circle. The diametre divides the circle into two symmetrical figures called semi-circles.

A chord of a circle is any straight line drawn across a circle, but starting and ending on the curve making the circle. A diametre is also a chord.

An arc of a circle is any piece of the curve which makes the circle.

Exercise 11a

1 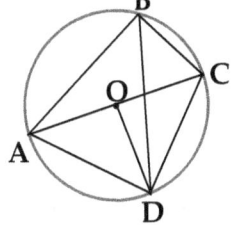 (a)   Given the circle below, identify and write down all the possible radii, diametres, chords and arcs you can obtain. For example, AB is a chord,

(b)   From the arcs obtained, state whether each of them is minor or major. For example arc AB (direct) is a minor arc, and AB (through C and D).

2 (a)   Given the following figure. join the points to form all the chords we can have. For example chord *ab*.

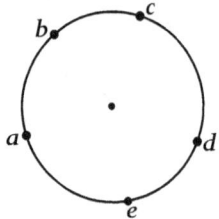

(b)   What is the highest possible number of chords formed.

11.1.2 Circle parts: *According to shape*

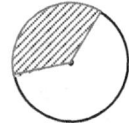
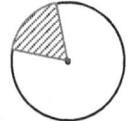

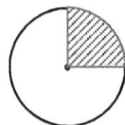

Sector   Minor sector   Major sector   Quadrant

(a quarter of a circle)

A sector of a circle is a shape enclosed between an arc and the two radii at the either ends of that arc.

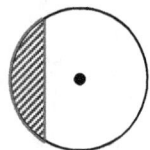
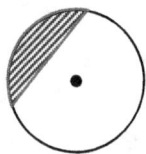
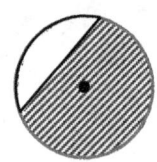

Segment   Minor segment   Major segment   Semi-circle

(a half of a circle)

A segment of a circle is the shape enclosed between a chord and one of the arcs joining the ends of that chord.

Exercise 11b

1

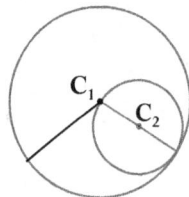

(a)   Given the figure on left shade the major sector of the big circle.

(b)   On another drawing of the same figure, shade the part of minor sector of the big circle which is not part of the small circle.

2 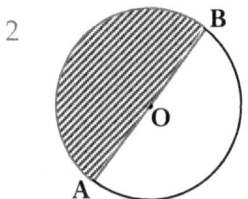 From the figure given, state whether the shaded part is a sector or a segment and explain why.

11.2 Drawing Circles

Circles can never be got accurately unless they are drawn with a specific method.

(a) When you place your foot heel fixed on ground and rotate your first toe, you obtain a circle.

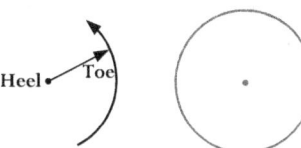

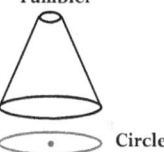

(b) We can also use circular objectives like circular tin lids, coins, plates, tumblers, etc. Cups are often used in shaping pan cakes.

The above methods of circle drawing are fixed, depending on the size the foot or circular object.

(c) A string fixed at a point (centre) is kept taut (pulled tight) with its other end fixed with a pencil. The pencil is rotated about the fixed point, drawing a circle.

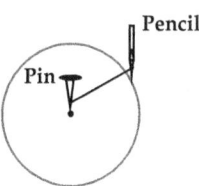

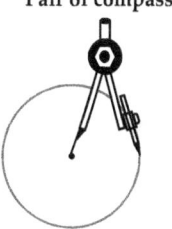

(d) All the above methods (a) to (c) are not convenient, therefore the best method in (d) is preferred. This uses an instrument known as the compasses.

11.3 Related circle

11.3.1 Concentric circles

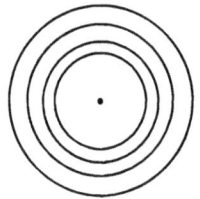

Concentric circles are two or more circles which have been drawn using the same centre.

11.3.2 Eccentric circles

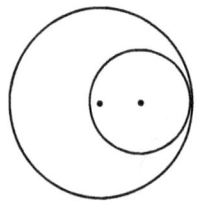

Eccentric circles are two or more circles which have been drawn using different centre, usually with one circle completely.

Activity 11.1

One circle has a radius of 7cms and another circle with radius equal to 4cms has its centre on the radius of the first, 2cm away from the centre of the first one. Draw these circles and state their relationship.

11.3.3 An annulus

An annulus is a shape like a ring which is formed by the shape enclosed between two concentric circles.

REVISION EXERCISE 11

1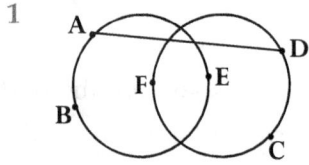

On the figure, join the points which form chords for both circles and label them. For example, AD forms chords for both circles as opposed to AB which only forms a chord for only one circle.

2 Three circles of equal radii are arranged evenly such that they exactly fit in a distance of 9cm. What is the radius of each circle?

3 Draw four concentric circles with radii, 3 cm, 5cm, 7cm and 8cm.

4 Draw a circle of 3.5cm radius. Using the same radius and at any point on the circle curve, draw another circle. At each of the two points where the two circles cross, draw the same radii circles as the first circles. Join all the four circle centres with straight lines.

5 One circle has a radius of 4cm and another smaller circle has a radius, 3cm. The centre of the smaller circle is located such that it is 1 cm away from the centre of the bigger circle along the radius. Draw the two circles, and state their relationship.

6 Draw four circles with radius, 2cm, 3cm, 5cm and 7cm eccentrically such that the external of each circle just touches the inside of the successive larger circle.

7 Draw two circles concentrically such that the smaller has a radius 3.5cm and the width of the annulus they form is 1.5cm. find the diameter of the larger circle.

8 Draw a straight line AB, 10cm long and divide it into equal parts of 2cm each. After every 2cm, draw a circle with centre on line AB with diameter 4cm such that no circle goes beyond the points A and B. What is the largest number of circles can you achieve?

12 NUMBER PATTERNS AND SEQUENCES

INTRODUCTION

A set of objects can be arranged in order following a given rule. Such an arrangement is a pattern. Patterns are characterized by repetition of the rule on the object or number in order to achieve the next. Each separate object or number in the arrangement is called a *term*.

12.1 Quantity and order

Terms of the number patterns form an order in which their terms values reduce or increase in a way.

12.1.1 Whole number

Whole numbers include: 0, 1, 2, 3, 4, 5, 6 . . . ; and can further be broken down to other number categories;

Natural (or counting) numbers:	1, 2, 3, 4, 5, 6 . . .
Odd numbers:	1, 3, 5, 7 . . .
Even numbers:	0, 2, 4, 6 . . .
Prime numbers:	2, 3, 5, 7 . . .
Composite numbers:	4, 6, 8, 9 . . .

All the above sets of numbers are formed according to specific rules.

Number category	*Rule (or trend of numbers' accumulation)*
Odd numbers	*Each number or term exceeds the next (or is less than the previous) by 2.*
Prime numbers	*number with no factors other than 1 and itself.*
Square numbers	*Multiplying a natural number by itself.*

Example 1

Fill in the missing numbers: 10, 12, 14, ___ , ___ , 20, ___ .

Solution

There are even numbers, so each exceeds the previous by 2.
So we have: 10, 12, 14, <u>16</u> , <u>18</u> , 20, <u>22</u> .

Example 2

Copy and complete: 16, 8, 4, ___ , 1 , ___ , ___ .

Solution

There are even numbers, so each exceeds the previous by 2.
We have: 16, 8, 4, <u>2</u> , 1 , $^1/_2$, $^1/_4$.
$$\div 2 \qquad \div 2 \quad \div 2$$

Example 3

Complete the following: 12, 9, 7, ___ , ___ , ___ .

Solution

We have: 12, 9, 7, <u>6</u> , <u>6</u> , <u>7</u> .
$$-3 \quad -2 \quad -1 \quad \pm 0 \quad +1$$

Exercise 12a

Copy and complete the following:

1   1 , 2, 4, 8, ___ , ___ , ___ , ___ .   2   1 , 3, 6, 10, ___ , ___ , ___ , ___ .

3   17, 16, 19, 18, ___ , ___ , ___ .   4   2, 3, 5, 7, ___ , ___ , 17, ___ .

5   1, 3, 5, ___ , 9, ___ , ___ , 15.   6   ___ , 78, 67, 56, ___ , ___ .

7   ___ , 7, ___ , 13, 16, 19, ___ .   8   ___ , ___ , ___ , -7, ___ , -1, 2, 5, 8

Activity 12.1

Study carefully and complete the following patterns with out calculations.
Explain the trend of answers in each case.

1				2		
	8 x 1	+ 1	= 9		9 x 1	= 9
	8 x 12	+ 2	= 98		9 x 12	= 108
	8 x 123	+ 3	= 987		9 x 123	= 1107
	8 x 1234	+ 4	= 9876		9 x 1234	= 11106
	8 x 12345	+ 5	= ___		9 x 12345	= 111105
	8 x 123456	+ 6	= ___		9 x 123456	= ___
	8 x 1234567	+ 7	= ___		9 x 1234567	= ___
	8 x 12345678	+ 8	= ___		9 x 12345678	= ___
					9 x 123456789	= ___

12.1.2 Object patterns

Objects can also be used to form patterns.

Example 1

Complete the following object patterns:

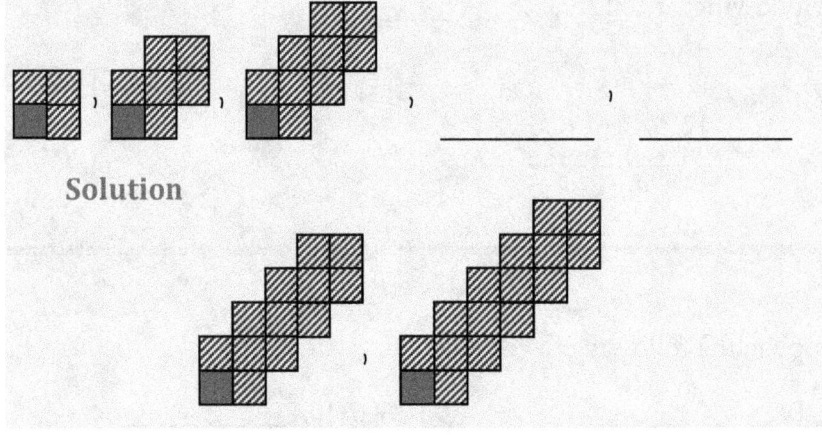

Solution

Exercise 12b

Draw the next two figures in each of the sequences.

1

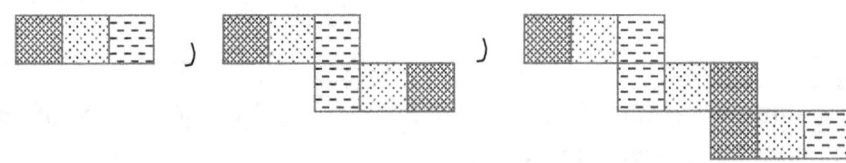

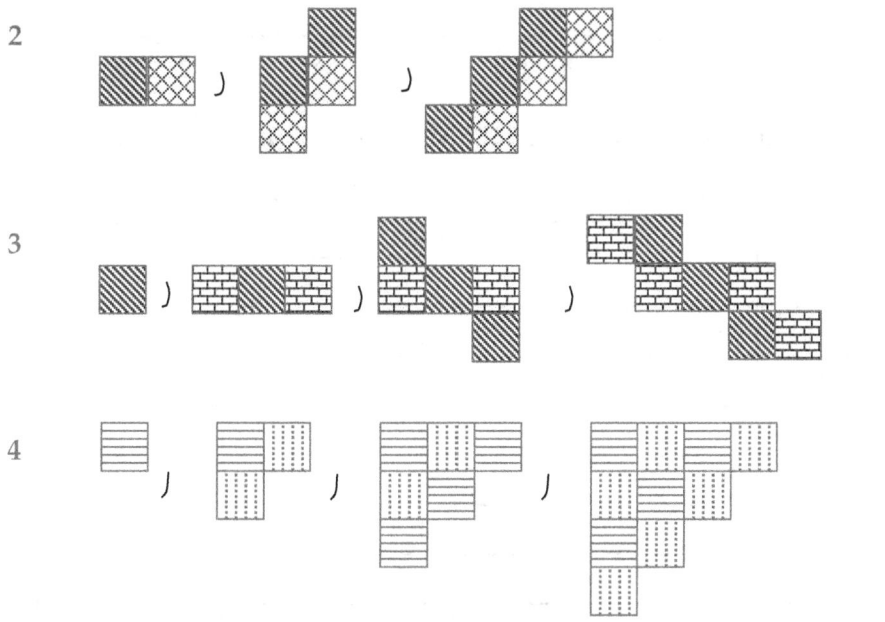

12.1.3 Divisibility test

In Chapter four (page 64), we looked at the divisibility tests and we found out trends to know whether a number is divisible by the other. This trend follows a certain pattern in each case.

Study the following tables carefully and complete them. Check the sums of digits to find out whether they are divisible by 3.

No.	Patterns (sum of digits)	Divisible by 3 ? YES	Divisible by 3 ? NO
01	0 + 1 = 1		√
02	0 + 2 = 2		√
03	0 + 3 = 3	√	
04	0 + 4 = 4		
05	0 + 5 = 5		
10	1 + 0 = 1		
11	1 + 1 = ---		
12	1 + 2 = ---		

No.	Patterns (sum of digits)	Divisible by 3 ? YES	Divisible by 3 ? NO
14	---		
15	---		
100	---		
102	---		
103	---		
104	---		
105	---		
1050	---		

Exercise 12c

Copy and complete the following tables. First state the rule in each case.

1

Number	Pattern (sum of digits)	Is the number divisible by 6 ?	
		Yes	No
1012	1 + 0 + 1 + 2 = 4		√
48114	4 + 8 + 1 + 1 + 4 = ?		
3816			
3314			
548163			
111111			
3036			
91200			

2

Number	Pattern (sum of digits)	Is the number divisible by 9 ?	
		Yes	No
3093	3 + 0 + 9 + 3 = 15		
41355	4 + 1+ 3 + 5 + 5 = ?		
11481			
92007			
41148			
26187			
120006			
342067			
111008			
99900			

3

Number	Pattern (sum of digits)	Is the number divisible by 6?	
		Yes	No
36102	3 + 6 + 1+ 0 + 2 = 12		
46011			
33812			
24288			
111111			
1006700			
1322217			
7800123			

12.2   Number sequences

We normally use number sequences to make Calendars. Using number sequences we can know what date it would be (or it was) after a given period of time. If today was Wednesday 30th October 2002, we can tell the day and date after 5 days. After 5 days the day and date will be Monday 4thJ November 2002.

If today was Wednesday 6th November 2002, the day and date after 16 days will be Friday 22nd November 2002. Check this from the following figure.

November 2002 Calendar

SUN	MON	TUE	WED	THU	FRI	SAT
					1	2
3	4	5	6	7	8	9
10	11	12	13	14	15	16
17	18	19	20	21	22	23
24	25	26	27	28	29	30

Fig. 121

In the rows, we realize that 1 is added for every day that passes. For example, from Sunday 3rd; we have 3rd, 4th, 5th, 6th, 7th, 8th and 9th. Therefore, the pattern as you move from the left to the right is; *"keep adding one"*.

The sequence involved is: 3, 4, 5, 6, 7,8, 9.

In the columns we realize that 7 is added for every week. For example, the dates of all the Sundays in the month are 3rd, 10th, 17th and 24th. Therefore, the pattern as you move from the top to the bottom is; *"keep adding seven"*. The sequence involved is: 3, 10, 17, 24.

Activity 12.2

Make a calendar for the month of January starting on Monday.
(a) What day will it be on 21st?
(b) What date will it be on the second Sunday in the month?
(c) On what day will the month end?

A *sequence* is therefore a set of numbers in which each number follows the last according to a uniform rule. The numbers forming a sequence are called *terms*. Always remember that the terms are usually separated with commas in a sequence.

12.2.1 Arithmetic sequences

A sequence obtained by adding the same number is called an *arithmetic sequence*. On a calendar we used the arithmetic sequences to determine the previous or future dates. Other examples of arithmetic sequences are odd and even numbers:

Odd = { 1, 3, 5, 7, 9, 11, 13, 15, 17, 19, ... }

Even = { 0, 2, 4, 6, 8, 10, 12, 14, 16, 18, ... }

Generally, if we have a term say, n the next term for both odd and even numbers is $(n + 2)$.

For example, if n = 1, then (1 + 2) = 3, so the next term after 1 is 3 for odd numbers. This therefore makes the difference between any two consecutive terms for both even and odd numbers equal to 2.

Check:-   Is 8 – 6 = 2 ?   *Yes*

Is 18 – 16 = 2 ?   *Yes*

Is 9 – 7 = 2 ?   *Yes*

Try to get the difference between any other consecutive even or odd numbers, what do you realize? You will always get a 2.

This 2 is called the *common difference*.

If we know the common difference, then we can generate as many terms as possible.

Example 1

Copy and complete the following sequence: 3, 6, 9, ___ , ___ , ___ .

Solution

The common difference is (6 - 3) = 3

So we have: 3 , 6 , 9 , 12 , 15 , 18
                    +3   +3   +3   +3   +3

Example 2

Copy and complete the following sequence: 25, 20, 15, ___ , ___ , ___.

Solution

The common difference is (20 - 25) = -5

Always remember that, subtracting is the same as adding a negative.

So, we have: 25 , 20 , 15 , 10 , 5 , 0
                    -5   -5   -5   -5   -5

Exercise 12d

Complete the following arithmetic sequences:

1   7, 8, 9, 10, ___ , ___ , ___ .

2   4, 7, 10, ___ , ___ , ___ .

3   1, 5, ___ , ___ , ___ , 21.

4   30, 25, 20, ___ , ___ , ___ .

5   2, 10, ___ , ___ , 34, ___ .

6   6, -3, 0, 3, ___ , ___ , ___ .

12.2.2   Geometric sequences

A sequence obtained by always multiplying the previous term by the same number (*common multiplier*) is called a *geometric sequence.*

In the promotion of a new product sale, coupons were made to be distributed and later a grand draw was made.

At the beginning the trend was;

- *Two people were given an original coupon each.*

- *Each of them wrote his name on the coupon then photocopied to get 2 copies and supplied them to other two people, retaining the original.*

- *Each person who received a coupon did the same thing. (that is, getting 2 photocopies and distribute them).*

The distribution of the coupons was as follows;

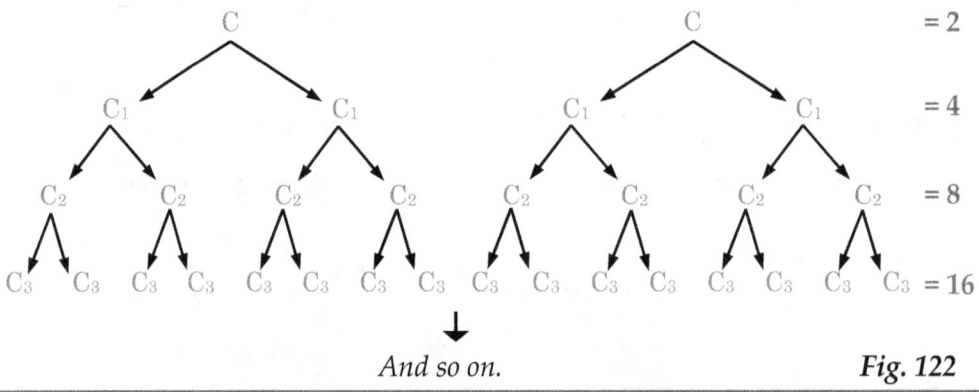

And so on.

Fig. 122

Where,

C represents the original coupon

C_i represents the first photocopy

C_{ii} represents the second photocopy

C_{iii} represents the third photocopy and so on.

This gives us the following geometric sequence: 2, 4, 8, 16 ...

x2 x2 x2

In this case 2 is the *common multiplier*.

Other geometric sequences may be;

4, 12, 36, 108 ...

x3 x3 x3

1, 6, 36, 216, 1296 ...

x6 x6 x6 x6

Activity 12.3

State the common multipliers and list the next 3 terms in each of the following sequences.

(a) 1, 5, 25, ___ , ___ , ___ .

(b) 1, 8, 64, 512, ___ , ___ , ___ .

(c) 3, 12, 48, 192, ___ , ___ , ___ .

Exercise 6c

List the missing terms in the following geometric sequences.

1 -3, -6, -12, ____ , ____ , ____ .

2 1, -5, 25, -125, ___ , ___ , ___ .

3 ___ , ____ , 24, 48, ____ , 192.

4 0.2, 0.6, ___ , 5.4, ___ , ____ .

5 ¼, ¾, ___ , ____ , ____ , $243/4$

6

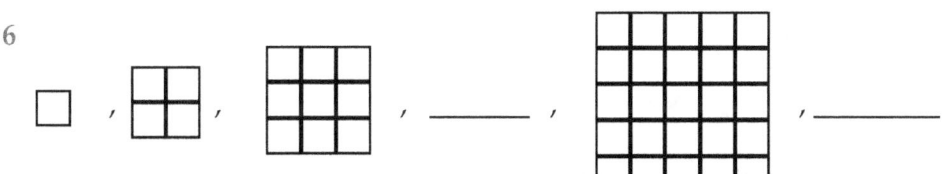

12.3   Other number patterns

12.3.1   Triangular numbers

Study the arrangements (patterns) of stars in the triangles in figure that follows.

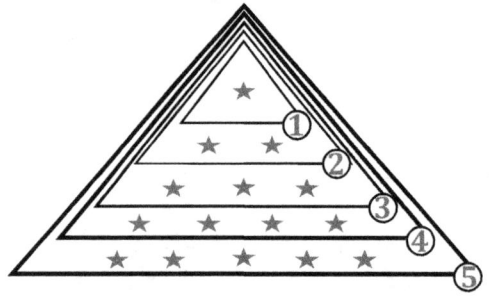

Triangle 1 has 1 star,

Triangle 2 has 3 stars,

Triangle 3 has 6 stars,

Triangle 4 has 10 stars,

Triangle 5 has 15 stars, etc. ...

Fig. 123

These triangular numbers are obtained as follows;

1	→	1
1 + 2	→	3
1 + 2 + 3	→	6
1 + 2 + 3 + 4	→	10
1 + 2 + 3 + 4 + 5	→	15
↓		↓
etc.		etc

Therefore, triangular numbers 1, 3, 6, 10, 15, ... are obtained by adding natural numbers continuously, forming the next triangular number each time an addition is performed.

Activity 12.4

(a) Continue the pattern above on the right up to the 10th triangular number.

(b) A number is 5 times the second triangular number,

   (i) What is the number

   (ii) State whether this number is triangular or not.

12.3.2 Rectangular numbers

To obtain rectangular numbers we shall consider numbers that form both rectangular and linear arrays.

A number 6 has both linear and rectangular arrays as follows;

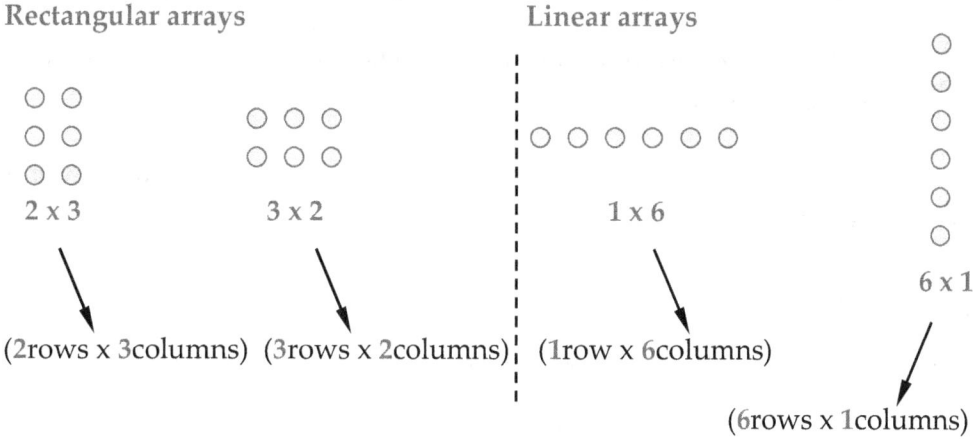

Fig. 12.4(a)   Fig. 12.4(b)

A *linear array* has the number of rows or columns or both equal to 1. A *rectangular array* has both the number of columns and rows more than one, except for the number 1 which can make both linear and rectangular arrays.

Study the arrays for the numbers 3, 4, 5 and 8 in the follows table.

Number	Rectangular array(s)	Linear array(s)	
3	None	o o o 1 x 3	or	o o o 3 x 1
4	2 x 2	o o o o 1 x 4	or	o o o o 4 x 1
5	None	o o o o o 1 x 5	or	o o o o o 5 x 1
8	2 x 4 or 4 x 2	o o o o o o o o 1 x 8	or	o o o o o o o o 8 x 1

Fig. 12.5

The array for the number 1, (i.e. 1 x 1) is both rectangular and linear

The numbers 4 and 8 form both linear and rectangular arrays, therefore they are rectangular numbers.

Whereas the numbers 3 and 5 form only linear arrays, therefore they are not rectangular numbers.

Activity 12.5

(a) Form all the rectangular arrays that can be obtained from the number 12.

(b) Try to form rectangular arrays for all prime numbers below 50. Make a comment for your answer.

(c) 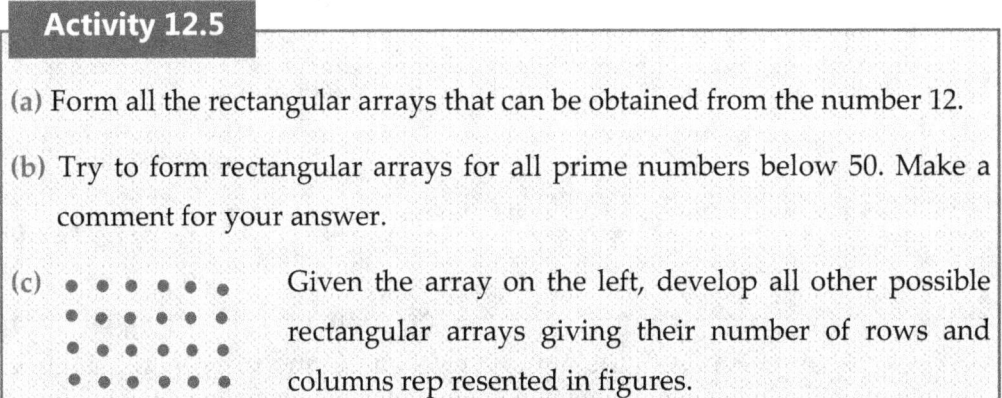 Given the array on the left, develop all other possible rectangular arrays giving their number of rows and columns rep resented in figures.

Example 1

Supposing you are given to cut a piece of paper 6units long and 2units wide into 1unit by 1unit pieces. How many small pieces of paper can you obtain from the big paper? If you arrange the obtained small pieces to form one big sheet of paper, show the different sizes which can be formed.

Solution

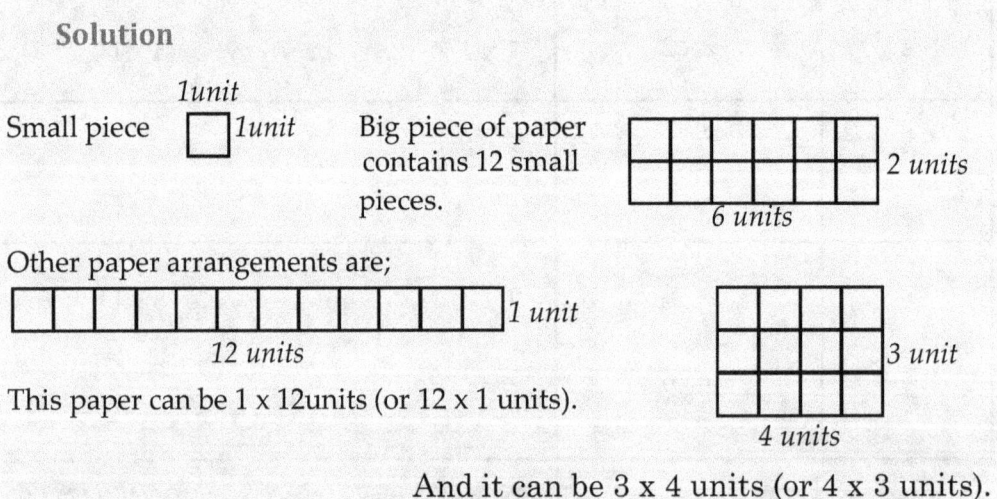

Other paper arrangements are;

This paper can be 1 x 12units (or 12 x 1 units).

And it can be 3 x 4 units (or 4 x 3 units).

Exercise 6d

1 The table below shows arrays of numbers, study them carefully and complete the table:

Number	Rectangular pattern	Arrangement	Naming	Is the number rectangular ?	
				Yes	No
1	o	1 x 1	One 1	√	
2	None	None	None		√
3	None	None	None		√
4	o o o o	2 x 2		√	
5	None	None	None		√
6	o o o o o o o o o o o o	2 x 3 3 x 2	Two 3s Three 2s	√	
7					
8					
9					
10					
11					
12					

Have you noticed that;

    (a) the rectangular arrangements reduce a number to its factors, and

    (b) all numbers are rectangular except the prime numbers?

2 A large sheet of paper 9dm by 8dm, was exactly divided into smaller sheets of paper of 3dm by 2dm by cutting with out wasting any piece.

    (a) Find how many small pieces can be obtained from the large piece.

    (b) Find the sizes of other pieces of paper which can be obtainable from all the cut small sizes, showing their different arrangements.

12.3.2 Square numbers

Given the pattern: 1, 4, 9, etc. you will obtain arrays such as;

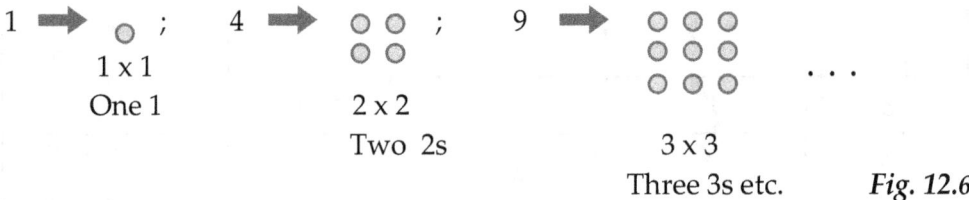

    Fig. 12.6

You will realize that for each array the number of rows equals to the number of columns. The number 9 is a square number with equal factors 3 and 3. The equal factors are called *square roots*. According to square patterns all square numbers are rectangular numbers.

Exercise 6e

1 Copy and complete;

    1J = 1 x 1 = 1

    2J = 2 x 2 = 4

    3J = 3 x 3 = 9

    4J = _____

    5J = _____

    6J = _____

    7J = _____

    8J = _____

    9^2 = _____

    10J = _____

2 Form the square arrays for the following numbers

    (i) 25   (ii) 36   (iii) 64

3 Given the following square numbers, form their other arrays (rectangular), apart from their linear or square arrays.

    (i) 36   (ii) 49   (iii) 81

4 State true or false;

    (i) Square numbers can not form linear arrays

    (ii) All square numbers can form rectangular arrays.

12.3.4 Relationship between odd and square numbers

Let us try to add odd numbers continuously and see what answers we obtain.

Odd numbers = {1, 3, 5, 7, 9, 11, 13, ...}.

On adding, to accumulate odd numbersJ sums we have;

1	= 1	Each time an odd number is
$1 + 3$	= 4	added in a cumulative manner,
$1 + 3 + 5$	= 9	the result is a square number.
$1 + 3 + 5 + 7$	= 16 etc.	

1 is odd and a square number; adding the next odd number 3, we obtain a square number 4, adding the next odd number 5, we obtain a square number 9, adding the next odd number 7, we obtain a square number 16, and so on.

Therefore the relationship between odd and square numbers is, each time a consecutive odd number is added to the previous, starting from 1, a square number is obtained.

So when we add odd numbers continuously, we obtain a square number each time we add the odd number.

That is,   1; $1 + 3 = 4$; $1 + 3 + 5 = 9$; $1 + 3 + 5 + 7 = 16$;   ...

Exercise 6f

1 Copy and complete the following patterns

1^2 =   1   =   1

2^2 =   4   = $1 + 3$

3^2 =   9   = $1 + 3 + 5$

4^2 =   16   = $1 + 3 + 5 + 7$

5^2 =

6^2 =

7^2 =

8^2 =

9^2 =

10^2 =

2 (a) Find the square number formed if seven consecutive odd numbers are added starting from 1.

   (b) How many consecutive odd numbers are added to form a square number 81? List them.

3 (a) Find the 2 consecutive odd numbers which are added to 64 to obtain 100.

   (b) Complete the following pattern
   $1 + 3 + 5 + ... + 13 + 15$
   ↓ ↓ ↓   ↓   ↓
   1 4 9 ...   .

12.3.5 Relationship between triangular and square numbers

Each time any two consecutive triangular numbers are added, a square number is obtained. For example, 1 and 3 are two consecutive triangular numbers and if they are added, a square number 4 is obtained.

Carefully study the square patterns and their triangular representation in the following figure.

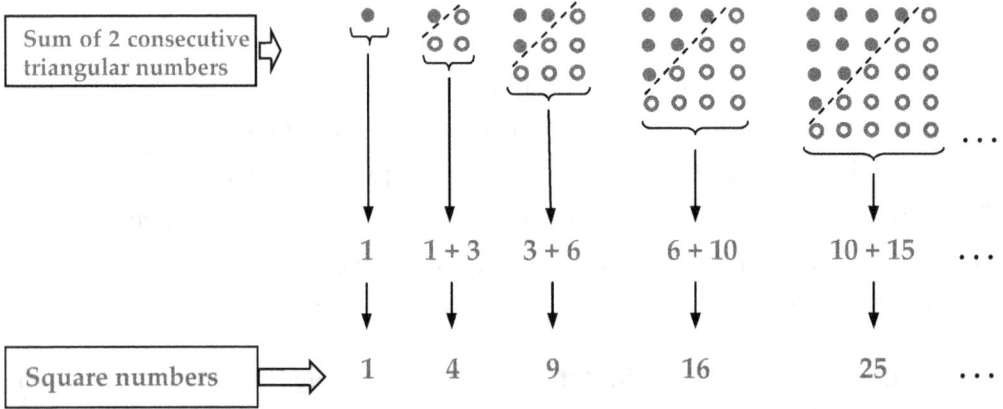

Fig. 12.5

Activity 12.6

Given the square numbers below find the pattern of the consecutive triangular numbers that were added to obtain these square numbers,
(i)   36   (ii)   225

12.3.6 Cubic numbers

For 64 = 4 x 4 x 4, the number 4 which we multiply by itself twice is called the *cube root*, and the number 64 is the *cubic number*. So the cube of 4 is 64 and the cube root of 64 is 4.

Cubic numbers are: 1, 8, 27, 64, 125, …

Exercise 6g

1 Study the following patterns carefully, copy and complete them.

$$1^3 = 1 \times 1 \times 1 = 1$$
$$2^3 = 2 \times 2 \times 2 = 8$$
$$3^3 = 3 \times 3 \times 3 = 27$$
$$4^3 = 4 \times 4 \times 4 = 64$$
$$5^3 = 5 \times 5 \times 5 = 125$$
$$6^3 =$$
$$7^3 =$$
$$8^3 =$$
$$9^3 =$$
$$10^3 = 10 \times 10 \times 10 = 1000$$

2 Workout the following cubes

(i) 12 x 12 x 12 (ii) 13 x 13 x13

(iii) 15 x 15 x 15 (iv) 20 x 20 x 20

3 What is the cube of,

(i) 18 (ii) 23

(iii) 125 (iv) 205

4 Find the cube roots for,

(i) 1000 (ii) 8000

(iii) 512 (iv) 1331

Activity 12.7

(a) Find a number (s) apart from 1, which is (are) both square and cubic.

(b) Explain how the relationship between a square and a cubic number can be obtained.

12.3.7 Finding subsets of a given set

In Chapter 1, we looked at how to generate and find the number of subsets for a given set. This can be generalized using a certain trend. From the table on page 12, chapter 1, let us pick columns 1 and 4 as follows;

Number of elements in the set (n)	Numbers of subsets (s)
0	1
1	2
2	4
3	8
4	16

The pattern for the number of sub sets is a geometric pattern, that is;

$$1, \quad 2, \quad 4, \quad 8, \quad 16 \ \ldots$$
$$\times 2 \quad \times 2 \quad \times 2 \quad \times 2$$

When we look at the values of the number of elements in a set and its corresponding number of subsets we realize a relationship between them as follows;

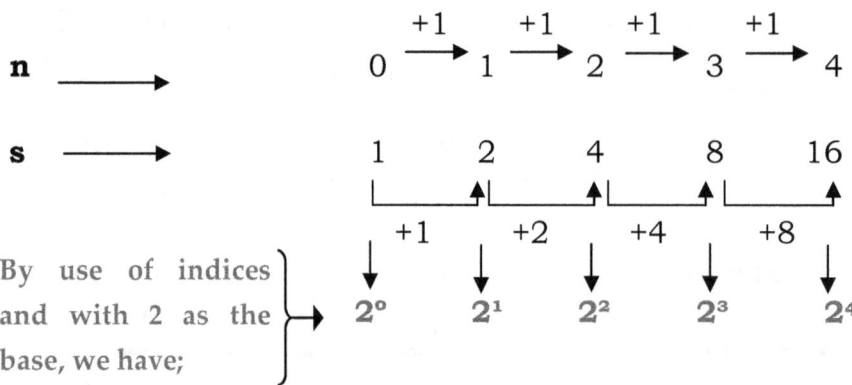

By use of indices and with 2 as the base, we have;

From the above pattern we realize that there is a relationship between the number of subsets, **s** and their respective powers in the indices with 2 as the base, that is;

Have you realized that the powers of 2 correspond to the n-values in each column?

Therefore, $s = 2^n$; generally, where s is the number of subsets and n is the number of elements in the given set.

Conclusively, the number of subsets for a particular set will be 2 to the power of the number of elements in that set.

In the following table study carefully the trends and relations in finding the number of sub sets using the number of elements the mother set has.

Copy and complete the following table below;

No. of elements in the set (n)	No. of subsets by formula	No. of subsets (s)
0	2^0	1
1	2^1	2
2		
3		
4		
5		
6		
7		
8		
9		
10		
⋮	⋮	⋮
n	2 x 2 x ... up to n times	2^n

Exercise 6h

1 The sum of three consecutive triangular numbers is 46. Find these triangular numbers.

2 Find the triangular number represented by: $1 + 2 + 3 + \ldots + 10$

3 What is the eighth triangular number?

4 List the triangular numbers less than 100 and are (i) Square numbers
(ii) Prime numbers

5 Which 2 consecutive triangular numbers do we add to obtain the square number 100?

6 List the consecutive odd numbers that are added to obtain the square number 100 ?

7 What is the fifth cubic number?

8 How many subsets does a set with 12 elements has?

REVISION EXERCISE 12

1 Given the digits 4, 6, 5 and 9, find all the possible numbers you can form using these digits which are divisible by: (i) 4 (ii) 11

2 If today is Wednesday 1st January 2003, what will the dates of all the Wednesdays of this month be?

3 Complete the following sequences

(i) -4, -3, $^3/_5$, ___ , ___ , -2 $^2/_5$, ___.

(ii) 3, ___ , ___ , $^3/_{64}$, $^3/_{256}$, ___.

---

4 Copy and Complete the following sequences:

(a)

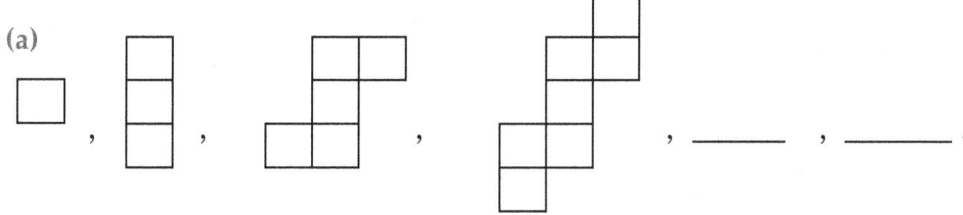

(b)

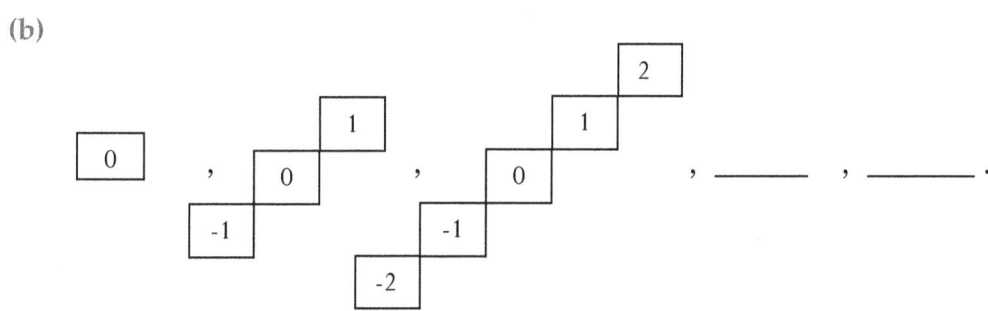

5   (a)   Find the numbers less than 100 which are both cubic and square numbers.

(b)   From the list you have made, find numbers having square roots which are cubic numbers.

APPROXIMATION AND ESTIMATION

INTRODUCTION

If I asked for your age, you may say, *'fifteen'*, although your actual age could be fifteen years, four months, three weeks and two days. In your answer therefore you will have approximated your age in years.

We can deal with a number of situations involving the applications of approximation and estimation.

13.1 Terminologies involved

Accuracy:
The accuracy of a number is the measure of how exact it is. If a group of five students measured a book 320mm long, we expect answers such as 321mm, 323mm, 320.5mm, 319mm or 319.5mm. It is not easy to get the exact answer. Realizing this, we find it necessary to give the obtained answer and just indicate its nearness to the actual length (accuracy). For example, the most accurate student will be that with the answer 319.5mm if stated to the nearest whole number (which gives 320mm) .

Approximation:
In a case where the exact number can not be obtained, the stated value close to it, is its approximation. $^4/_{11}$ as a decimal gives a recurring (never-ending) decimal 0.363636 . . . ; however, this can be approximated as 0.4, 0.36, 0.364 and so on. Each of these values is close to $^4/_{11}$, however, their accuracies vary.

Estimation:
When we obtain an approximation depending on judgement and not by calculating, measuring or carrying out any process that may give a more accurate answer, then it is said to be an estimation.

A news reporter may reach a scene where rioters are being dispersed by anti-riot police unit, and states his title as *"200 dispersed by police – South Street"* When the actual counting is done, you may find that the number of rioters was 182, 221 or 242. In this case the 200 is only an estimation of 182, 221 or 242.

13.2 Rounding off numbers

Before rounding off any given number it is necessary to know which measure of accuracy you are to use.

13.2.1 Measures of rounding off numbers

While rounding off numbers, we normally measure the magnitude of accuracy according to;

(i)   the *number of decimal places*, always presented as " to . . . decimal places (d.p)". " . . . " represents the number of decimal places. For example, to 3 decimal places or 3 d.p in short.

(ii)   the *place values*, always represented as "to the nearest . . .". " . . . " represents the place value. For example, to the nearest hundredths.

(iii)   *Significant figures (sgf)*, always expressing the relative importance of the digits in the number. In this case the number of digits of importance.

13.2.2 How to round off numbers

To round off the number in each of the three categories above, we;

1   Identify the digit in question

2   Look at the next digit on the right of that in (i) above.

3   If it is less than 5, then maintain the digit in question and drop all those on its right.

4   If it is 5 or more, then raise the digit in question by 1 and make all those on its right.

Example 1

Write the decimal 394.8621, (i) correct to 1 place of decimal.

(ii) correct to 2 places of decimal.

Solution

(i)

One place of decimal (Circle the number).

3 9 4 . ⑧ 6 2 1

Next digit to the right is greater than 5.

Add 1 to 8 (*in place of one decimal digit*) to obtain 9 and ignore the digits 6, 2 and 1 on the right of 8.   So, 394.8621 is equals to 394.9(to 1 d.p)

(ii)

Two places of decimal (Circle the number).

3 9 4 . 8 ⑥ 2 1

Next digit to the right is less than 5.

Maintain 6 (*in place of two decimal digits*) and ignore the digits 2 and 1.
So, 394.8621 is equal to 394.86 (to 2 d.ps).

Example 2

Write the decimal 73.8574, (i) to the nearest whole number

(ii) to the nearest hundredths

Solution

(i)

Nearest whole number.

7 ③ . 8 5 7 4

Digit to the right of the required place value.

The immediate next digit to the right of the nearest whole is greater than 5, therefore, we add 1 to 3 and ignore all the other digits to the right.

So, 73.8574, is approximately equal to 74 (to the nearest whole number)

(ii)

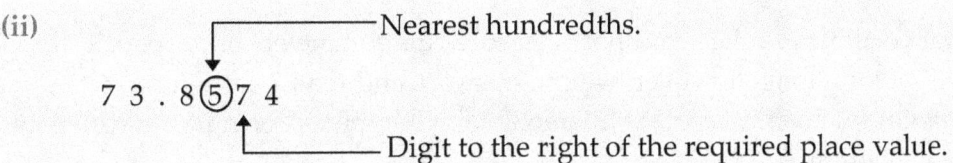

The immediate next digit to the right of the nearest hundredths is greater than 5, therefore, we add 1 to 5 and ignore all the other digits to the right.

So, 73.8574, is approximately equal to 73.86 (to the nearest hundredths)

In chapter 6, we looked at place values of decimal numbers, so if you have a problem in understanding them, you can revisit this chapter, pages 111 to 113.

Exercise 13a

Write the following decimals to the nearest whole number.

1	304.245	2	4.986	3	406.9998
4	0.012	5	10.009		

Write the following decimals correct to one decimal place.

6	410.00129	7	300.549	8	1804.39
9	306.398	10	0.108116		

Write the decimal 1404.9821;

11	correct to the nearest hundreds
12	correct to the nearest hundredths
13	correct to the nearest tenths
14	to the nearest whole number
15	to one place of decimal

13.2.3 Significant figures (Whole numbers)

Most numbers are based on measurement, and all measurements are not exact, so there is need to use only those digits that are meaningful (or significant) in recording the numbers obtained from any measurement.

Your ordinary ruler can not measure the diameter of a pencil to be 7.435218mm long; however, we can measure and read 7.4mm from it. So, it is meaningful to measure the diameter of your pencil using your ruler and obtain 7.4mm.

We need to note that;

1. All whole numbers (except zero) are significant.

For example;   1, 2, 3, 4, 5, 6, 7, 8 and 9 are significant figures.

0 is not significant

2. A zero in front of a digit or group of digits is non significant.

For example; The zeros in 023, 0004 and 01456 are non significant.

3. A zero behind a digit or group of digits is non significant.

For example; The zeros in 1200, 4540 and 873000 are non significant

4. A zero between a two digits or a group of digits is significant.

For example; The zeros in 1208, 23004 and 7106 are significant.

From the notes above;

1.   023 has 2 significant figures

2.   01456 has 4 significant figures (1 is the first s.f)

3.   1200 has 2 significant figures

4.   87000 has 2 significant figures (7 is the second s.f)

5.   1208 has 4 significant figures

6.   50300 has 3 significant figures (The last two zeros are not significant)

Example 1

Write the following numbers to 3 significant figures (i) 2476   (ii) 34089

<u>Solution</u>

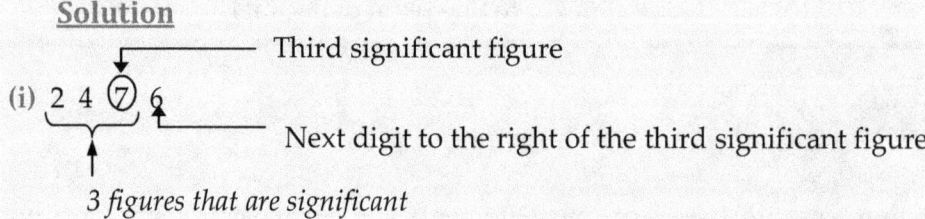

In this case 6 is greater than 5, so we replace it with zero to keep the place value and add 1 to 7 to get 9. So, 2476 is equal to 2480 (to 3 s.f).

(ii)

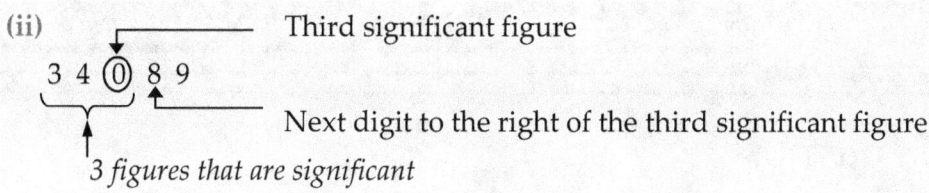

Third significant figure

3 4 ⓪ 8 9

Next digit to the right of the third significant figure

3 figures that are significant

In this case 8 is greater than 5, so we replace it and 9 with zeros to keep the place values and add 1 to 0 to get 1. So, 34089 is equal to 34100 (to 3 s.f).

Exercise 13b

Write the following to 2 significant figures

1	201	2	3275	3	4416
4	4198	5	3307		

Write the following to 3 significant figures

6	87408	7	10518	8	3319		
9	0.03921	10	66.456	11	20086		
12	30415	13	66105	14	5159		
15	9087098						

REVISION EXERCISE 13

Find the number of significant figures for the numbers in each case below;

1   50098   2   5113   3   100200   4   3408   5   40050200

Express the number 324.847019

6   to the nearest whole number   7   correct to 2 places of decimal

8   to the nearest thousandths   9   to 1 significant figure

10   to 3 significant figures

14 COMMERCIAL AND HOUSE HOLD ARITHMETIC

INTRODUCTION

At home we buy some products, pay for services and utilities. While doing these we engage in a number of commercial activities at home, shops and at service or utility centres.

We buy commodities such as food stuffs, house hold items, electronic equipment, cutlery, furniture, sanitary items and many other house hold goods.

We pay for services like education, medical care, telecommunication and many more. We also pay for utilities like water and electricity.

14.1 Budgeting and Costing

Before you buy commodities, you need to write down what you want, how much of it and at what cost. We can variously tabulate the budget or costing information for ease of analysis.

14.1.1 Budgeting for item

When we are preparing to go out to markets, shops or at any other shopping centres, we first make a budget for what to buy.

The major components of a budget involve:

1 *the item number,*
2 *description,*
3 *quantity,*
4 *rate and*
5 *amount for each.*

This can then provide a total at the end.

Back to school stationery budget

No.	Description	Quantity	Rate (Ush)	Amount (Ush)
1.	Exercise books (96 pages)	20 pieces	500	10,000
2.	Pens	2 dozens	600	1,200
3.	Pencils	1 dozens	300	300
4.	Ruled papers	1 ream	8,000	8,000
5.	Duplicating papers	1 ream	12,000	12,000
6.	Geometric sets	1 piece	3,500	3,500
7.	File folders	6 pieces	2,500	15,000
8.	Crayons	1 dozen	2,500	2,500
	TOTAL			52,500

Exercise 14*a*

Make tabulated budgets for the following sets of items

1   (a) 3 shirts at Ush. 3,500 each

    (b) A calculator at Ush. 15,000 each

    (c) A radio at Ush. 21,000 each

    (d) 2 pairs of shoes at Ush. 25,000 each

    (e) 4 pairs of shorts at Ush. 8,000 each

    (f) 5 T-shirts at Ush. 1,800 each

2   (a) 3 hens at Ush. 10,800 each

    (b) 6 basins worth Ush. 15,000 each

    (c) 3 watches worth Ush. 25,500 each

    (d) 3 dozens of plates, each plate at Ush. 300

    (e) 4 pairs of shorts at Ush. 8,000 each

    (f) 9 tables at Ush. 82,000 each

14.1.2 Price lists

There are a number of price lists, such as menus, shopping lists, price tags, etc.

RESTAURANT MENU			
Soft drinks		**Snacks**	
1. Soda 500ml	Ush. 1,500	1. Chips plain	Ush. 2,500
2. Soda 300ml	Ush. 600	2. Chips + Chicken	Ush. 4,500
3. Splash 100ml	Ush. 1,000	3. Chips + Liver	Ush. 4,000
4. Splash 500ml	Ush. 3,500	4. Chips + Beef	Ush. 4,000
5. Passion juice 500ml	Ush. 1,000	5. Egg roll	Ush. 800
6. Yogurt 500ml	Ush. 800	6. Sausages 1 pair	Ush. 1,000
7. Yogurt 250ml	Ush. 350	7. Chaps	Ush. 1,000
8. Mineral water 500ml	Ush. 500	8. Bread slice	Ush. 400
9. Mineral water 1000ml	Ush. 1,200	9. Burger	Ush. 2,000
10. UHT Milk 500ml	Ush. 4,500	10. Boiled egg	Ush. 500

Exercise 14*b*

From the menu given above, find the total cost for each of the set of items given below.

1 2 bottles of 300ml soda, 3 mineral water bottles (500ml), chips and liver and a pair of sausages.

2 4 glasses of passion fruit juice (500ml), 2 glasses of yogurt (500ml), and 1 glass (250ml), 4 pieces of chaps and 1 egg roll.

3 Mineral water (500ml), Chips plain, Egg roll, 2 pieces of chaps and 4 plates of chips and beef.

4 Passion fruit juice (500ml), 3 packs of 500ml UHT milk, 5 bread slices, 5 burgers and 2 plates of chips and chicken.

SHOPPING LIST			
Item	**Price**	**Item**	**Price**
1. Shoes	Ush. 54,000	8. Hand bags	Ush. 35,000
2. Shirts	Ush. 5,000	9. Gaggles	Ush. 2,500
3. Belts	Ush. 2,500	10. Socks a pair	Ush. 2,500
4. Skirts	Ush. 3,000	11. Pants (ladies)	Ush. 3,000
5. Blouses	Ush. 3,000	12. Pants (men)	Ush. 5,000
6. Trousers	Ush. 15,000	13. Suit case	Ush. 45,000
7. Suits	Ush. 75,000	14. Neck tie	Ush. 3,000

Exercise 14*c*

From the shopping list provided above, copy and complete the table below;

No.	Description	Quantity	Rate (Ush)	Amount (Ush)
1.	Socks	3 pairs	_____	_____
2.	WomenJs pants	_____	3000	9000
3.	Neckties	6 pcs	3000	_____
4.	Skirts	11 pcs	_____	_____
5.	Suits	_____	_____	105,000
6.	Trousers	_____	_____	75,000
7.	Shoes	_____	54,000	162,000
8.	Gaggles	_____	_____	87,500
	TOTAL			_____

14.2 Buying and Selling

We always buy goods or services from those who sell them. The sellers can manufacture or produce the goods or services they sell. In another way they buy the goods or services for resale. By doing this, they are looking for a profit margin, which is actually the money they make on selling the goods or providing the services.

14.2.1 Prices, Profit and Loss

There are two levels of prices as attached to the businesses.

Cost Price (CP):- This is the price of making or purchasing an article. If the article was bought for resale, and not made, the cost price can also be a purchase price. The cost price may also involve transport, taxes, storage, handling and other expenses of the article on top of its purchase price.

Selling Price (SP):- This is the price at which an article is offered for sale. This is the price at which the consumer takes an article.

Prices may vary depending on the level of buying or selling. For example, if a good is manufactured we can have:

Manufacturer	*Business Person*	*Consumer*
Only selling	Buying and selling	Only buying
Cost price	Purchasing and Selling price	Selling price

When a business person buys goods for resale, he/she intends to make a profit, although some times they incur losses.

If SP > CP, then a profit is realized

If SP < CP, then a loss is inccured.

Therefore, mathematically,   Profit = SP − CP

Loss = CP − SP

Example 1

A trader bought a bar of soap at sh. 720 and sold it at sh. 1,020. Find the profit he made.

Solution

Profit = SP − CP   But SP = sh. 1,020 and CP = sh. 720

Profit = sh. (1,020 − 720)

Profit = sh. 300.

Exercise 14*d*

Given the cost price (CP) and selling price (SP) of articles as follows; calculate and state the profit or loss made in each case.

1 CP = sh. 1,650; SP = sh. 3,400   4 CP = sh. 1,650; SP = sh. 2,100

2 CP = sh. 1,520; SP = sh. 1,415   5 CP = sh. 2,250; SP = sh. 1,750

3 CP = sh. 2,300; SP = sh. 3,050

Find the selling price for each of the following items;

6 C.P = sh. 1,250, gain sh. 150   9 C.P = sh. 1,470, loss sh. 360

7 C.P = sh. 7,900, gain sh. 1,020   10 C.P = sh. 720, no gain or loss

8 C.P = sh. 1,250, loss sh. 250

Find the cost price for each of the following items;

11 S.P = 1,050, gain sh. 320   14 S.P = 3,600, loss sh. 1,200

12 S.P = 3,220, gain sh. 1,030   15 S.P = 2,450, no gain or loss

13 S.P = 1,075, loss sh. 225

14.2.2 More on profit and loss

Profit and loss can also be presented as percentages of the cost price of commodities.

Example 1

After selling a radio at sh. 45,000; a retailer realized that he had made a 20% loss. Find the price at which he purchased the radio.

Solution

Percentage loss = $\dfrac{\text{Loss}}{\text{CP}} \times 100$ But loss = CP − SP

By substitution we have;

Percentage loss $\quad = \dfrac{CP - SP}{CP} \times 100$

$\qquad 20 \qquad\quad = \dfrac{(CP - 45,000)}{100CP}$

Then solve to find CP, the Cost Price.

$\qquad\qquad 20CP \;=\; 100CP - 4,500,000$

$\qquad\qquad {}^{-}80CP \;=\; {}^{-}4,500,000$

$\qquad\qquad \Rightarrow CP \;=\; 56,250$

Therefore, the Radio was purchased at sh. 56,250 by the retailer.

Exercise 14*e*

Find the selling price in numbers 1 to 5.

1 C.P = sh. 6,400; Profit = 25%   4 C.P = sh. 10,650; Loss = 15%

2 C.P = sh. 5,600; Profit = 12%   5 C.P = sh. 12,400; Loss = 10%

3 C.P = sh. 1,800; Loss = 5%

Find the cost price in numbers 6 to 10.

6 S.P = sh. 1,800; Profit = 12%   9 S.P = sh. 18,500; Loss = 7½%

7 S.P = sh. 5,400; Profit = 3%   10 S.P = sh. 18,500; Loss = 8%

8 S.P = sh. 12,000; Profit = 12%

Find the percentage profit or loss in numbers 11 to 15.

11 S.P = sh. 1,800; C.P = sh. 1,500   14 C.P = sh. 12,300; Loss = sh. 800

12 C.P = sh. 2,000; Profit = sh. 400   15 S.P = sh. 18,000; Profit = sh. 1,200

13 S.P = sh. 6,400; Loss = sh. 400

16 I purchased a book worth sh. 12,000 at a discount of 15%. I then sold it at sh. 11,000.

(a) Find my cost price of the item

(b) What was my percentage profit on the item?

17 While at a trade fair an exhibitor reduced an item he had purchased at sh. 50,000 from sh. 75,000 to sh. 45,000. A retailer purchased it and resold it

at a percentage profit of 12%.

(a) What percentage loss did the exhibitor make?

(b) Find the percentage reduction of the item made at the trade fair.

(c) At what price did the retailer sell the item?

14.3 Bills and tariffs

Normally when we use utilities such as water and electricity, we pay for them. Other services we pay for include telephone calls, transport, rent, postage and mailing, and so on. Some of these services are pre-paid whereas others are post paid.

A pre-paid service is one which you pay for before using it. Pre-paid service examples include airtime load cards, postage stamps, insurance policy stickers, and so on.

A post paid service is one which you pay for after using it. For example your monthly bills for water and electricity show how much of the utilities you have consumed. A bill is the document detailing how much is due for the customer to pay for the service or utility.

A tariff is a charge or system of charge used to bill the customer for services.

14.3.1 Water bills

A unit for measuring water consumed is a cubic metre (m^3). A cubic metre is equivalent to 1000 litres of water. In Uganda, the National Water and Sewerage Corporation is the major body responsible for the management of water and sewerage.

Water is charged as follows;
1 Number of units at a fixed price each.
2 Service charge per month.

Value Added Tax (VAT) as a percentage charge over the sum of items 1 and 2 above. VAT is normally at 18%.

Example 1

Alice received her November bill with the following readings;

    Current reading: 10721

    Previous reading: 10839

Evaluate to find the value of her November bill if each unit costs sh. 2730 and the service fee per month is sh. 2,000.

Solution

Units consumed	$= (10839 - 10721)$ m^3 $= 118$ m^3
118 m^3 at sh. 2730	= sh. 322,140
Service fee charge	= sh. 2,000
Subtotal	= sh. 324,140
ADD: 18% VAT	= sh. 58,345.2
Total bill due for payment	**sh. 382,485.2**

Example 2

In February 2012, I received my January bill with a balance of sh. 106,000 carried from December 2011. In January I used 32m^3 of water at a rate of sh. 2,500 and a service charge of sh. 2,100 for that month. Calculate the amount for payment according to my bill.

Solution

Balance brought forward		= sh. 106,000
32m^3 at sh. 2,500 each	= sh. 80,000	
Service charge for January	= sh. 2,100	
Subtotal	= sh. 82,100	
ADD: 18% VAT	= sh. 14,778	
January total		= sh. 96,878
Final total		**sh. 202,878**

Exercise 14f

1 In the month of March, John used 90m^3 of water. If the balance carried from

February was sh. 31,000 and that the monthly service fee is sh. 3,200 and VAT of 18%, how much is due for him to pay including all charges?

2 My water metre reading now is 30112 units and was 30003 units when it was last read. I had no balance carried forward. If the service fee is sh. 2,150 per month and VAT is rated at 18%, how much am I supposed to pay the water board?

3 I last paid for my water bill in July and my account was over paid by sh. 81,000. I did not pay for the water for the next two months. I received a bill combining August and September. Its current reading was 130645 units from the last reading of 130125 units for July. If the service fee for each month is sh. 2,600 and VAT charged at 18%.

(a) Find the amount due for the two months.

(b) What is the net amount I am supposed to pay to the board?

4 If I used 20,500 liters of water in three months,

(a) how many units were these?

(b) if the board charges a service fee of sh. 2100 per month and 18% VAT, how much would I be required to pay for the water used?

5 I am supposed to pay sh. 23,600 for the water I used last month. If the service fee for a month is sh. 2,000 and VAT charge is 18%, how many litres of water did I use that month if each unit is at sh. 1,800?

14.3.2 Electricity bills

The unit used to measure electric current consumed or used is the kilowatt-hour (kWh). 1 kWh is equivalent to putting an electrical gadget of 1 kW (or 1000 watts) on power for 1 hour.

For example;

- A 1000 watts cooker on electricity for 1 hour, consumes 1 unit (1 kWh).

- A 2000 watts electric kettle on electricity for half an hour (or 30minutes), consumes 1 unit (1 kWh) of electric power.

- A 100watts bulb takes 10 hours on electric power to consume 1 unit (1 kWh) of electric power.

Example 1

How many units of power are consumed by an electric kettle rated at 3,000 watts for 3 hours?

Solution

Total power consumed	= 3,000 watts x 3 hours
	= 9,000 watt hours
	But 1000watt hours = 1 kWh
So, 9,000 watts	= 9 kWh Therefore it consumes 9 units.

Example 2

How many 25watts bulbs do I need to consume a unit of power with in 1 hour?

Solution

Number of bulbs	=	$\dfrac{1000 \text{ watts}}{25 \text{ watts}}$
	=	40 bulbs

Therefore, 40 – 25watts bulbs will consume a unit of power in 1 hour.

When the electricity board bills the power consumers, it rates then according their consumption. The first 15 units may be charged at a different rate from the next afterwards. On this, a service fee is charged and a tax (Value Added Tax) is topped up, normally at 18%.

Example 3

Ali used 35 units of power. In a particular month the first 15 units are charged at sh. 120, and the rest thereafter at sh. 320 each. A service fee of   sh. 3,500 per month and an 18% VAT rate is charged.
(i) Find the total amount Ali is supposed to pay for that month.
(ii) If his previous balance was sh. 504,000, what total would be due to him?

Solution

Cost for first 15 minutes	15 x 120	=	sh.	1,800
Cost for rest of the units	20 x 320	=	sh.	6,400
Subtotal 1		=	sh.	8,200
Add: Service fee		=	sh.	3,500
Subtotal 2		=	sh.	11,700
Add: VAT 18%		=	sh.	2,106
Total due for payment		=	**sh. 13,806**	

Exercise 14g

1 How many 40watts radios can consume 160watts at a time?

2 How many 20 watts bulbs can consume a unit of power with in 30 minutes

3 How many 2000watts kettles do consume 20 units of power in 4 hours?

4 I consumed 133 units of power in a month. That month a service fee of sh. 3,450 was charged, sh.120 for the first 17units, sh. 420 for the rest of the units and 18% VAT. How much money do I have to pay for that month?

Find the units of power consumed by each of the following electric gadgets in a given period of time it works,

5 A 2000watt kettles for 4 hours.

6 A 45watt bulb for 10 hours.

7 Five 20watts phone chargers for 60 hours.

8 A 2500watts cooker for 15 minutes.

9 A hot plate with two plates, one 2000watts and the other 3000watts working for 30 minutes and 1 hour respectively.

10 I used 125 units of power in a month. The first 15 units are charged at sh.320 each and the rest of the units thereafter are at sh. 520 each. If the service fee of sh.3500 per month and 18%VAT charged, find the
(i) total amount due for me to pay.
(ii) total amount due for me to pay, if I had a previous outstanding debt of sh. 220,000.

Copy and complete the following table by adding the column of total bill due.

	Units consumed	Months of consumption	Service fee per month	Cost for 1st 15 units	Rest of units cost	VAT
11	108	2	Sh. 2,005	Sh. 180	Sh. 320	18%
12	112	1	Sh. 1,008	Sh. 250	Sh. 400	12%
13	1,002	4	Sh. 1,250	Sh. 100	Sh. 350	18%
14	548	4	Sh. 3,200	Sh. 120	Sh. 400	15%
15	800	3	Sh. 3,000	Sh. 220	Sh. 450	18%
16	1,200	5	Sh. 2,500	Sh. 150	Sh. 420	18%

14.3.3 Telephone tariffs

Service providers charge their services according to the time a customer spends on air. Airtime cards with unique computerized secret access codes are used.

Their tariffs are based on seconds or minutes denominations alongside other special packages. A customer decides to pay per second or per minute.

If the service provider charges per minute, then every unit of the less than a minute is taken to be a minute. For example, if you are on a profile which charges per minute then 2 minutes 20 seconds will be considered 3 minutes. Even 2 seconds on this profile are considered a minute.

The table below can be used to find charges for the various service providers.

CHARGES PER MINUTE		To				
		Airtel	MTN	UTL	Warid	Orange
From	Airtel	Sh. 277	Sh. 500	Sh. 300	Sh. 300	Sh. 350
	MTN	Sh. 500	Sh. 300	Sh. 350	Sh. 350	Sh. 400
	UTL	Sh. 300	Sh. 400	Sh. 250	Sh. 300	Sh. 350
	Warid	Sh. 400	Sh. 450	Sh. 350	Sh. 199	Sh. 350
	Orange	Sh. 350	Sh. 350	Sh. 300	Sh. 300	Sh. 300

If the charge is per 15 seconds, then any value below 15 seconds is considered to be 15 seconds. For example, if you use 18 seconds, you will be charged for 30 seconds.

Example 1

A telephone service provider charges sh. 9 for every second when on profile P and charges sh. 330 per minute for profile Q.

(a) Which of the two profiles is generally cheaper than the other?

(b) If I use 1 minute and 40 seconds, how much do I have to pay on profile Q?

(c) How much money would I spend if I talked for 2 minutes 20 seconds on profile P?

(d) If I need to call for 2 ½ minutes, which profile would be cheaper for me?

Solution

(a)   Profile Q

(b)   40 seconds are equivalent to 1 minute on profile Q,
        so 1 minutes 40 seconds represents 2 minutes.

        Payment   = sh. 330 x 2 minutes

                    = sh. 660

(c)   2 minutes 20 seconds equals [(2 x 60) + 20] = 140 seconds.

        Payment on profile P = sh. 9 x 140 seconds

                            = sh. 1,260

(d)   Profile P   Profile Q

        2 ½ minutes gives;   2 ½ minutes gives;

        (2.5 x 60) seconds = 150 seconds   3 minutes

Pay: sh. 9 x 150 seconds   sh. 330 x 3 minutes

        = sh. 1,350   = sh. 990

Therefore, profile Q would be cheaper for me.

Try to find out the various charges as associated with service providers.

Exercise 14*h*

Using the tariff chart given on page 226 for the different service providers answer the following questions.

1 How much money do I need to call for 7 minutes from

    (i) Warid to UTL

    (ii) Orange to MTN

    (iii) Warid to Warid

    (iv) Airtel to Orange

2 If I wanted to make a call of 5minutes from one service provider to another,

    (i) Which combination can I make to achieve the lowest cost possible?

    (ii) How much will I spend?

3 Which call combination will I require to be charged the highest amount to call for 5 minutes? How will it be ?

4 If my service provider is Airtel and I loaded sh. 5000, then called MTN for 3 minutes, Airtel for 5 minutes and Warid for 2 minutes 45 seconds. How much money was my balance?

5 If my service provider is Warid, how much money do I need to call 2 ½ minutes on MTN, 7 minutes on UTL and 3 minutes on Orange?

A telephone service provider has three (3) profiles A, B and C. Customers on profile A are charged sh. 9 per second. Customers on profile B are charged sh. 100 for every 15 seconds. Customers on profile C are charged sh. 320 per minute. Using the above information, answer the following questions;

6 If a customer on profile A calls for 46 seconds, how much is he charged?

7 If a customer on profile B calls for 40 seconds, how much is he charged?

8 If a customer on profile C calls for 184 seconds, how much is he charged?

9 How much more does a customer pay to call for a minute if he is on profile

A instead of being on profile B.

10 What is the best profile do I need to call for the following times and how much will it be in each case?

(i) 12 seconds   (ii) 3 seconds   (iii) 10 minutes   (iv) 2 ½ minutes

14.3.4 Other tariffs

There are many other tariffs involved in our daily lives such as in transport fares, postage, announcements, advertisements, telegrams, and so on.

Example 1

To pass an announcement over a radio the management charges sh. 2,500 for the first 15 words and sh. 20 for any extra word there after. How much will the following announcement cost?

TO ALL EXECUTIVE COMMITTEE MEMBERS OF THE GREAT LIONS CLUB UGANDA

YOU ARE ALL DULY INFORMED TO ATTEND YOUR 13TH ANNUAL GENERAL MEETING DUE TO TAKE PLACE ON SURTADAY THE 22ND OF DECEMBER 2012 AT NEW PALACE HALL – MASAKA AT 10:30 A.M.

PLEASE ENDAVOUR TO ATTEND IN PERSON – MATTERS TO BE DISCUSSED ARE VERY CRUCIAL.

ANNOUNCED:

DAVID KATO CHAIRPERSON EXECUTIVE COMMITTEE.

Solution

First 15 words	= sh. 2,500
Next 46 words at sh. 20 each	= sh. 920
Total cost	**sh. 3,420**

Example 2

To send a parcel to the UK, it requires me to have it weighed and charged postage according to the weight. My payment should be made in terms of buying and sticking on the parcel postage stamps of an equivalent postage charge.

A parcel 20gms or less is charged sh. 1,250 and thereafter each extra gram is charged sh. 21.

(i) How much would it cost to send a parcel which weighs 35gms?

(ii) If the postage stamps to be used are in the sh. 5, sh. 10, sh. 20, sh. 50, sh. 100 and sh. 500 denominations, what is the minimum number of stamps required and their denominations?

Solution

(i) Cost for 20gms   = sh. 1,250
     Cost for more 15gms at sh. 21 each   = sh.   315
     Total cost   **sh. 1,565**

(ii) 6 stamps: sh. 500 + sh. 500 + sh. 500 + sh. 50 + sh. 10 + sh. 5

Exercise 14*i*

1 To post a parcel weighing 180gms or less to Indonesia, it requires sh. 5,500. Thereafter every additional 10gms are charged sh. 1,200. How much money do I need to send 4 parcels of weight 125gms, 220gms, 245gms and 188gms?

A bill board designer charges her customers as follows:

Every ¼ square metre of material	sh. 5,200
Full colour graphic	sh. 24,500
Upto 2 colour graphic	sh. 18,500
Every word	sh. 1,250

2 If I intend to design my advert on a ¾ square metre board, containing two full colour graphics and 15 words, how much would I pay to the graphic designer?

3 Ali made a 3m² bill board containing a 3 colour graphic and 32 words, how much was he charged by the graphic designer?

Below are charges for an Automated Teller Machine:

Cash deposit	sh. 100
Cash withdraw	sh. 200
Balance inquiry	sh. 150
Mini statement	sh. 350

4 If in a week I made 2 deposits, 3 cash withdrawals and a balance check, how much would I be charged?

5 How much would be charged for 7 cash deposits, 2 balances checks and a mini statement request?

REVISION EXERCISE 14

1 An item was purchased at sh. 58,240 after a discount of 20%. It was then sold at sh. 84,000.
   (a) Find the original cost price of the item.
   (b) What was the percentage profit on the item?

2 A manufacturer reduced an item price from sh. 54,000 to sh. 32,000. A wholesaler purchased it and resold it to a retailer at a percentage profit of 15%.
   (a) What percentage loss did the manufacturer make?
   (b) Find the percentage reduction of the item made by the manufacturer.
   (c) At what price did the wholesaler sell off the item?

3 I am supposed to pay sh. 23,600 for the water I used last month. If the

service fee for a month is sh. 2,000 and VAT charge is 18%, how many litres of water did I use that month if each unit costs sh. 1,800?

Copy and complete the table below.

Water consumed in one month:						
Previous reading	Current reading	Units used	Cost per unit	Service fee per month	VAT rate	Amount due
4 13601	13626	_____	Sh.3,200	Sh.3,100	18%	_____
5 _____	01928	36	Sh.2,100	Sh.2,500	18%	_____
6 08580	08614	_____	Sh.4,000	Sh.2,000	17%	_____

From the data below (questions 7 – 10) compute the total amount due for payment of electricity in each case for 1 month.

7 Used 108 units, sh. 250 for the first 15 units, sh. 400 for the rest of the units, sh.3,000 service fee per month and an 18% VAT charge.

8 Used 320 units, sh. 180 for the first 15 units, sh. 550 for the rest of the units, sh. 2,500 service fee per month and an 18% VAT charge.

9 Used 80 units, sh. 120 for the first 15 units, sh. 350 for the rest of the units, sh. 5000 service fee per month and 17% VAT charge.

10 Used 600 units, sh. 310 for the first 15 units, sh. 640 for the rest of the units, sh. 5100 service fee per month and 12% VAT charge.

BEARING AND SCALE DRAWING

INTRODUCTION

Bearing is an angle measured starting from the North direction towards the clockwise. A bearing is presented by 3 digits.

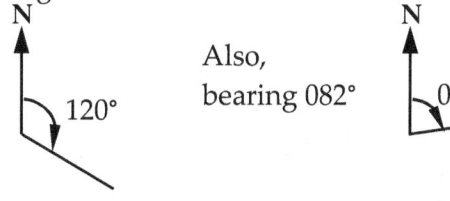

For example, the angle represented on the right is a clockwise measure of turn about point O, through 120°.

Also, bearing 082°

15.1 The Compass

In bearing we normally use compass directions and angles between them. Study the following compass directions carefully.

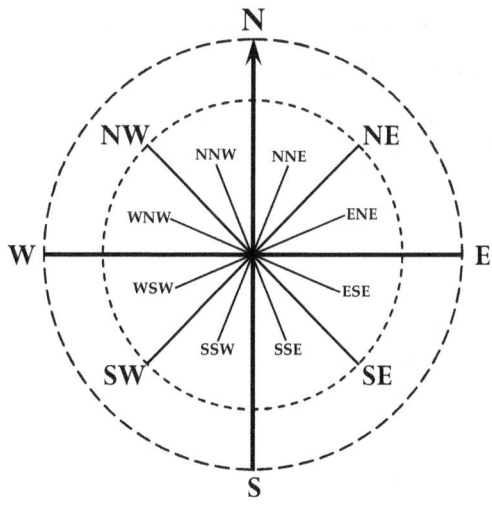

For the 16 directions as shown above, the angle between any two consecutive directions is 22½°.

15.1.1 Compass directions

The bearing of point P is 067°. This can also be represented as N67°E.

You should always remember that to move from the North back the North you go through 360°. So, revise you geometry for angles at a point to help you derive angles which are not given.

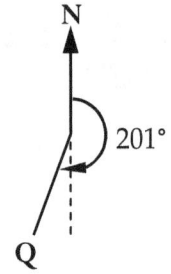

The bearing of point Q is obtained as:

$$(180° + 21°) = 201°.$$

Therefore, the bearing of point Q is 201°. That is point Q is 201° from North through a clockwise turn. This can also be represented as S21°W

Study the following table carefully.

Direction	Abbreviation	Bearing	Sketch
North	N	000°	
East	E	090°	
South	S	180°	
West	W	270°	
North East	NE	045°	
South East	SE	_____	
South West	SW	_____	
North West	NW	_____	
N50°E	-	050°	
N35°W	-	325°	
S65°E	-	_____	
S70°E	-	_____	

Activity 15.1

1 Represent the following directions on a sketch. For example;

128° ⇨   and   S52°E ⇨

2 Copy and complete the table above.

15.1.2 Opposites of compass directions

To easily compute directions, it is important we also know the opposites of directions on a compass, eg. Opposite of North is South, opposite of NW is SE.

Example 1

Find the opposites of the bearings: (i) 055°   (ii) 320°

Solution

(i) 055°   ⇨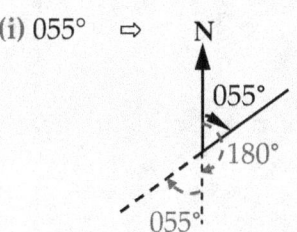

(ii) 320°   ⇨   N

∴ the opposite of 055° is:

(180° + 55°) = 235°

∴ the opposite of 320° is:

(320° - 180°) = 140°

Always remember that;

1 For a bearing between 0° and 180° you add 180° to the given bearing to obtain its opposite.

2 For a bearing between 180° and 360° you subtract 180° from the given bearing to obtain its opposite.

Exercise 15a

Following is a table for directions, bearings and their opposites. Copy and complete the table.

No.	Direction	Bearing	Opposites	
			Direction	Bearing
1	SW	225°	NE	045°
2	N54°E	_____	_____	234°
3	_____	138°	N42°E	_____
4	_____	_____	NNW	_____
5	_____	267°	_____	093°
6	S65°W	_____	N65°E	_____

15.2 Scale drawing

In this case, we represent and interpret the directions and angles by graphical means. We can sketch a situation and draw it accurately, then measure off the required dimensions.

Example 1

Asubira moved 3 km South, then 4 km East. In what direction is he finally from his starting point?

Solution

Use a _sketch_ as a guide   _Scale drawing_

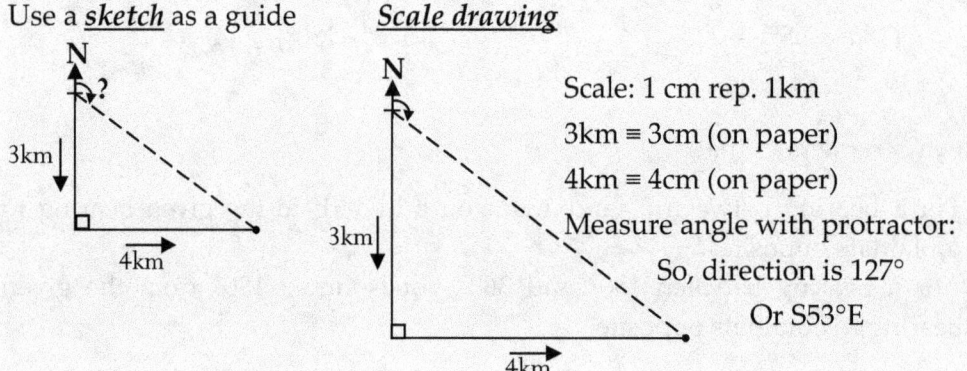

Scale: 1 cm rep. 1km
3km ≡ 3cm (on paper)
4km ≡ 4cm (on paper)
Measure angle with protractor:
So, direction is 127°
Or S53°E

Example 2

Kabete is starting at a point P and moves 5 km East and then he moves 12 km North to point Q. What is the

(a) shortest distance Kabete is away from his start point?
(b) bearing of Q from P?
(c) bearing of P from Q?
(d) relationship between the bearings of Q from P and P from Q?

Solution

Make sure you read and understand the interpretation of the question. Then make a sketch to guide you in your final scale drawing and measurements.

Sketch   *Scale drawing*

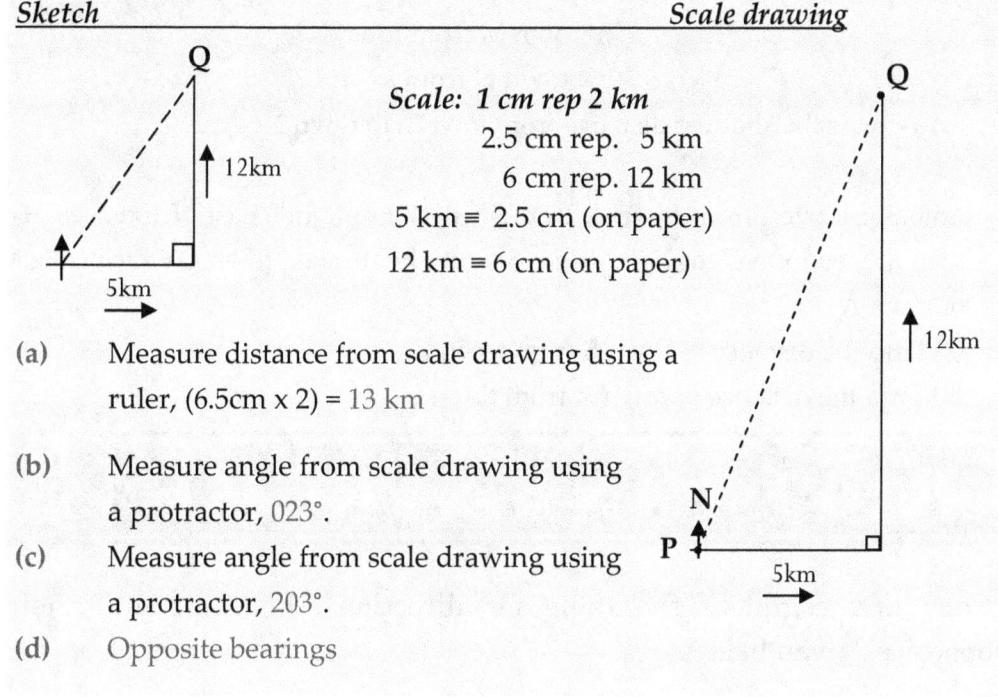

Scale: 1 cm rep 2 km
2.5 cm rep. 5 km
6 cm rep. 12 km
5 km ≡ 2.5 cm (on paper)
12 km ≡ 6 cm (on paper)

(a)   Measure distance from scale drawing using a ruler, (6.5cm x 2) = 13 km

(b)   Measure angle from scale drawing using a protractor, 023°.

(c)   Measure angle from scale drawing using a protractor, 203°.

(d)   Opposite bearings

Exercise 15*b*

1 I started moving from point A and continued to point B 5 km South East of A. I then turned to the North West direction and covered 5 km.

    (a) What is my final position?

    (b) Through how many degrees did I turn at point B?

    (c) What is the bearing of point B from A?

2 Town Q is 12 km North West of town P and town R is 15 km South East of town Q.

    (a) What is the distance between towns P and R?

    (b) What is the bearing of: (i) town P from town Q?

                                (ii) town P from town R?

3 From town L, I started moving towards the South West direction up to town M 11 km away from town L. at town M, I turned through 270° clockwise and continued to town N which is 17 m away from town M.

    (a) What is the bearing of: (i) town L from town M?

(ii) town M from town N?

(iii) town N from town L?

(b) What is the shortest distance from town L to town N?

4 Adibaku moved from town A to town B 17 km South East of town A. He then moved in the Northern direction until he reached town C which is East of town A.

(a) Find the distance of town B from town C.

(b) Find the distance of town A from town C.

REVISION EXERCISE 15

Copy and complete the table for directions, bearings and their opposites given below.

No.	Direction	Bearing	Opposites	
			Direction	Bearing
1	NW	315°	SE	135°
2	N62°E	_____	_____	242°
3	_____	105°	N75°W	_____
4	_____	_____	SSE	_____
5	_____	228°	_____	048°
6	S55°E	_____	N55°W	_____

7 Two cross roads intersect at 90°. Mujuzi and Katamba started their journeys from the road junction and each of them took a different road. After an hour, Mujuzi had covered 8 km and Katamba had covered 15 km.

(a) If the road Mujuzi took was continuing in the eastern direction;

(i) What are the possible road directions Katamba took?

(ii) What was the bearing of Katamba from Mujuzi after an hour of traveling?

(b) What was the shortest distance between Mujuzi and Katamba after an hour of travelling?

8 From town A, I moved in the direction S50°E and covered 7 km up to town B. From town B, I moved in the East for 3 km up to town C. From town C, I moved to town D 12 km away at a bearing of 030°.

(a) Find the bearing of: (i) town C from D

(ii) town B from D

(iii) town C from A

(iv) town D from A

(b) Find the shortest distance: (i) from town A to town C

(ii) from town B to town D

16 PROPERTIES OF GEOMETRICAL FIGURES

INTRODUCTION

From the concept of geometry, we have considered plane geometry which involves 2-dimensional figures and solid geometry involving 3-dimensional figures. Their properties and applications are mainly tackled.

16.1 Angles

When you are sent to pick cups from the cupboard, you have to open it, pick the cups and then close it after. As you open the door for the cupboard, it turns about the hinge. The difference in the positions of the door before and after opening it describes an angle about the hinge.

Similarly when you open your clipboard or your exercise book its cover describes an angle about the hinge. When two line segments meet at a point (vertex) they describe an angle.

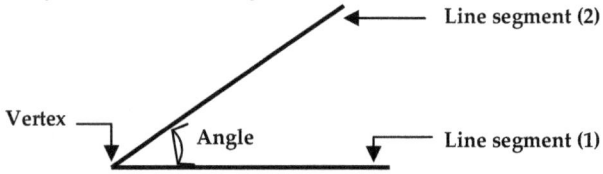

From the figure above, line segment (2) moved a circular distance away from line segment (1) rotating about the vertex, v. Therefore, an angle is a measure of turn. The unit for angle measure is a degree, for example 30 degrees, denoted 30°.

8.1.1 Types of angles

(a) Acute Angles: These are angles which are greater than 0º and less than 90°, that is, $0° < \theta < 90°$.

E.g. $\sphericalangle \theta = 63°, 54°, 20°, 89°$ etc.
Give other examples of Acute Angles.

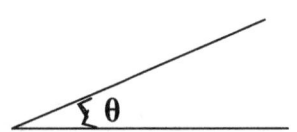

(b) Right Angles: These are angles with size exactly equal to 90°.
                       That is, $\angle\theta = 90°$.

(c) Obtuse Angles: These are angles which are greater than 90º and are less
                       than 180°, that is, **90° < θ < 180°**.

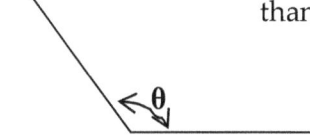

E.g. $\angle\theta = 162°$, 172°, 105°, etc.
Give other examples of obtuse angles.

(d) Reflex angles: These are angles which are greater than 180º and are
                       less than 360°, that is, **180° < θ < 360°**.

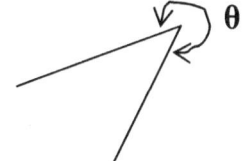

E.g. $\angle\theta = 274°$, 200°, 321°, etc.
Give other examples of obtuse angles.

Exercise 16a

State the types for each of the following angles according to their sizes.

    1 41°   2 231°   3 92°   4 113°   5   105°

    6 146°   7 38°   8 99°   9 120°   10 242°

8.1.2 Other categories of angles

(a) Adjacent angles: Adjacent angles are
angles which have a common vertex
and a common side but with no
common interior points.

So, $\angle\alpha$ and $\angle\beta$ are adjacent.
Or, $\angle ABD$ is adjacent to $\angle CBD$.

(b) Congruent angles: These are angles of a particular measure, similar to one
another and they have the same shape and size. They could share a vertex
or not. In a case where they share a vertex, then they are also adjacent.

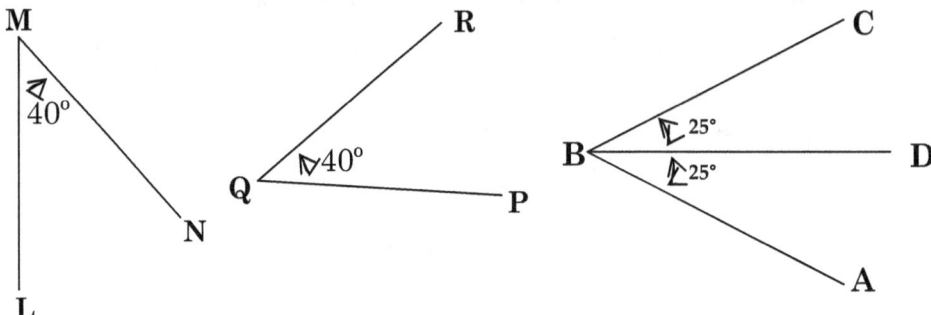

∢LMN and ∢PQR are congruent angles. So, ∢LMN = ∢PQR = 40°.

∢ABD = ∢CBD = 25°. So, ∢ABD is congruent to ∢CBD, and since the two angles share a vertex, they therefore qualify to be adjacent.

(c) *Complementary angles:* Are angles which when added their sum is equal to 90°. For example, if ∢σ = 60° and ∢α = 30°, then ∢σ and ∢α are said to be complementary angles, since ∢σ + ∢α = 90°. In any case, the condition for ∢σ and ∢α to be complementary, they must both be acute angles.

Activity 16.1

1 List 5 pairs of complementary angles.

2 If two complementary angles are congruent, what value should each be?

(d) *Supplementary angles:* These are angles which when added they sum up to 180°, that is, ∢β + ∢θ = 180°. For example, if ∢β = 123° and ∢θ = 57°, then ∢β and ∢θ are supplementary angles.

Activity 16.2

Can we have the two Supplementary angles which are both obtuse? If not why?

(e) *Angles on a straight line:* Angles on a straight line add up to 180° altogether. For example if we have;

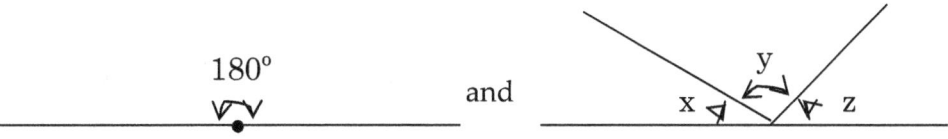

180°

and

Therefore, ∢x + ∢y + ∢z = 180°

Activity 16.3

Do you think you can find the value of z if you know the values of x and y? Explain your answer.

(f) Angles at a point: Angles at a point add up to 360°. That is,

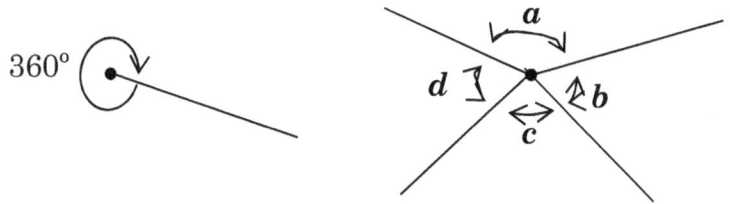

360°

Given angles at point, then; ∢a + ∢b + ∢c + ∢d = 360°.

Exercise 16b

1 Give the type of the following angles, (i) 63° (ii) 29° (iii) 90° (iv) 104° (v) 273°

2 Given the pairs of the angles that follow, state their categories.
(i) 45° and 45° (ii) 63° and 27° (iii) 90° and 90°
(iv) 120° and 60° (v) 33° and 147°.

3 (a) If the sum of angles 23°, 160° and x° form a straight line, find the size of angle x.
(b) There are 5 angles at a point: 35°, 104°, 121°, y° and 93°. Find the value of y.

4 (a) Given the figure next page, what is the category of the angles ∢RQA and ∢PQA?

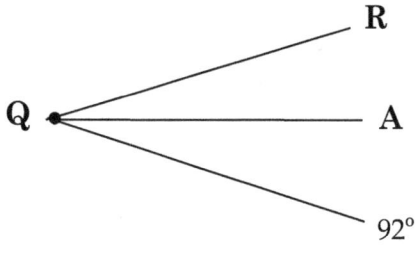

(b) If angle RQA = angle PQA , is it true that ⫟RQA and ⫟PQA are congruent angles? Explain.

5 Workout the angles represented by letters.

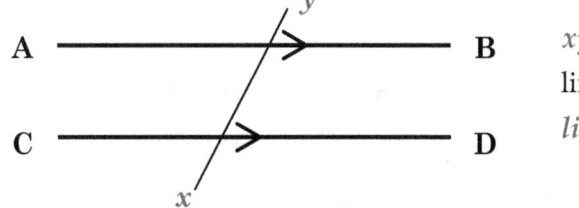

16.2 Parallel lines

Parallel lines are those lines which do not meet but follow the same direction. Examples of parallel lines are railway line, opposite edges of a ruler, wires on electric poles, etc.

xy – is a line cutting the parallel lines and is called a *transverse line.*

AB is parallel to CD, that is; AB ∥ CD. Parallel lines are shown by the arrows as follows;

or   or   depending on the various parallel lines existing in a particular instance, e.g. in the figure below the sides with the same number of arrows are parallel to each other.

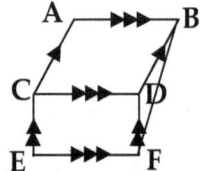

In the figure ABCDEF, AB ∥ CD, CD ∥ EF, AC ∥ BD, CE ∥ DF and the line BF is not parallel to any line in the figure.

16.2.1 Parallel lines and angles

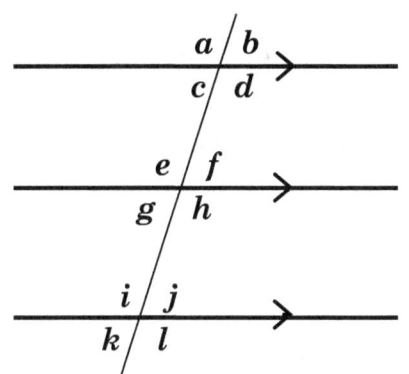

Consider the parallel lines chart to learn the angles involved when parallel lines are crossed by a transverse line.

Carefully study the chart and search for all possible relationships the angles made by a transverse line can form.

(a) *Corresponding angles:* Corresponding angles are equal. From the figure, angles *b* and *f* are corresponding angles; therefore, ∢b = ∢ f. Other corresponding angles are, ∢ b and ∢ j , ∢ c and ∢ g, ∢ a and ∢ e, etc. Comparing with the given pairs of corresponding angles to obtain all the other corresponding angles in the figure. If we know angle b then we can find angle *f* and *j*, since corresponding angles are equal. Thus we have; ∢ b = ∢ f = ∢ j.

(b) *Alternate angles:* Alternate angles are equal. From the figure we have ∢ c and ∢ f alternating therefore, ∢ c = ∢ f. Other alternating angles include ∢ j and ∢ c, ∢ f and ∢ k, ∢ g and ∢ j, etc. Study the figure clearly as you compare with the given pairs of alternate angles to obtain all the other alternate angles in the figure.

(c) *Vertically opposite angles:* Vertically opposite angles are equal. From the figure we have ∢ e and ∢ h, therefore, ∢ e = ∢ h. Other vertically opposite angles include ∢ i and ∢ l , ∢ a and ∢ d, etc. Study the figure carefully to obtain the other pairs of vertically opposite angles. If we know ∢ i, we can easily know the value of ∢ l since they are vertically opposite, that is, ∢ i = ∢ l.

(d) *Co-interior angles:* Co-interior angles add up to 180º. From the figure, ∢ d and ∢ f are co-interior angles, therefore ∢ d + ∢ f = 180º. Other pairs which are co-interior are ∢ g and ∢ i, etc.

Study the figure carefully and state other co-interior angles in the figure.

(e) *Co-exterior angle:* Co-exterior angles add up to 180°. From the figure, ∢ b and ∢ h are co-exterior angles, therefore ∢ b + ∢ h = 180°. The other pairs which are co-exterior in the figure are ∢ a and ∢ k. Study the figure clearly and find all the other co-exterior angles.

Example 1

Given the figure that follows, find the values of angles represented by letters and give reasons.

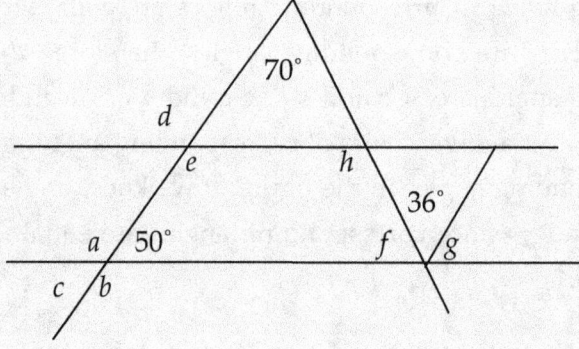

Solution

(i) ∢ a; Angles at a straight line add up to 180°

∴ ∢ a + 50° = 180°

∢ a = (180° − 50°)

⇒ ∢ a = 130°

(ii) ∢ b is vertically opposite to ∢ a,

∴ ∢ b = ∢ a

⇒ ∢ b = 130°

(iii) ∢ c is vertically opposite to the angle equal to 50°.

⇒ ∢ c = 50°.

(iv) ∢ d is corresponding to ∢ a.

Therefore, ∢ d = ∢ a

⇒ ∢ d = 130°.

(v) ∢ e is alternate to ∢ a, therefore ∢ e = ∢ a

∢ e = 130°

(vii) ∢ f; from the triangle we have three angles, i.e., 50, 70 and ∢ f. for a triangle, the interior angle sum is 180°.

So, ∢ f + 50° + 70° = 180

∢ f + 120° = 180

∢ f = (180° − 120°)

∢ f = 60°

(vii) ∢ g; angles on a straight line add up to 180°

Therefore, ∢ g + 36° + f = 180°

∢ g + 36° + 60° = 180°

∢ g + 96° = 180°

∢ g + 96° = 180°

∢ g = (180° − 96°)

∢ g = 84°

(viii) ∢ h, is a co-interior angle with ∢ f.

∢ h + ∢ f = 180°

∢ h + 60° = 180°

∢ h = (180° − 60°)

∢ h = 120°

Exercise 16c

1

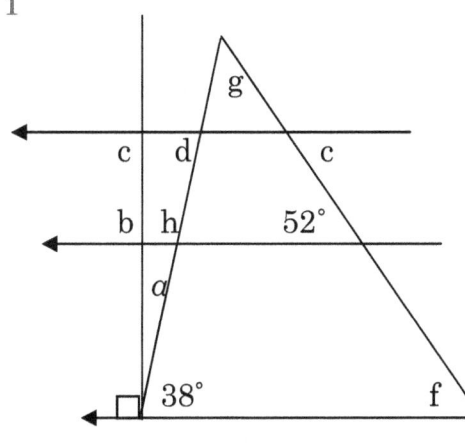

(a) In the figure above find the sizes of angles labeled by letters and give reasons for your answers.

(b) How are angles c and f related?

(c) Write expression relating angles a and h.

2

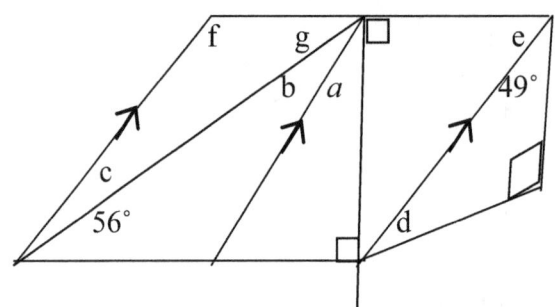

(a) From the figure given on left, find the angles labeled by letters and give reasons for your answers.

(b) Write an expression relating angles a, b and g.

(c) How are angles e and f related.

16.3 Polygons

Polygons are plane figures constructed by joining line segments at their end points. The line segments should not cross and the figure must be closed. Study the following figures carefully.

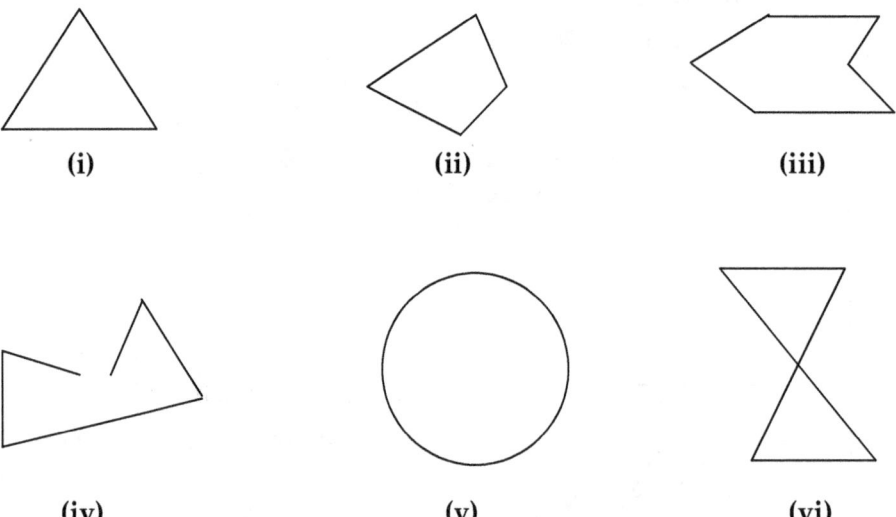

The figures (i), (ii) and (iii) are polygons. The figures (iv), (v),(vi) are not polygons because; figure (iv) is not closed, (v) has a curved boundary other than line segments and lastly (vi) has crossed line segments.

16.3.1 Convex and concave polygons

If each of the internal angles of a polygonal figure is less than 180°, then it is said to be a convex polygon. If any internal angle of a polygonal figure is greater than 180°, then it is said to be a concave polygon.

Study the following polygons and ensure that you can differentiate the convex from concave polygons. In this case strictly study the size of internal angles.

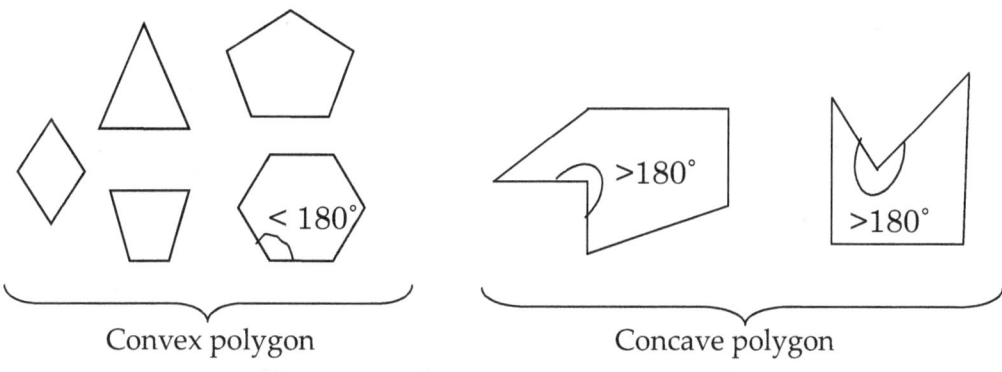

Convex polygon   Concave polygon

Exercise 16d

State whether each of the following polygons are convex or concave.

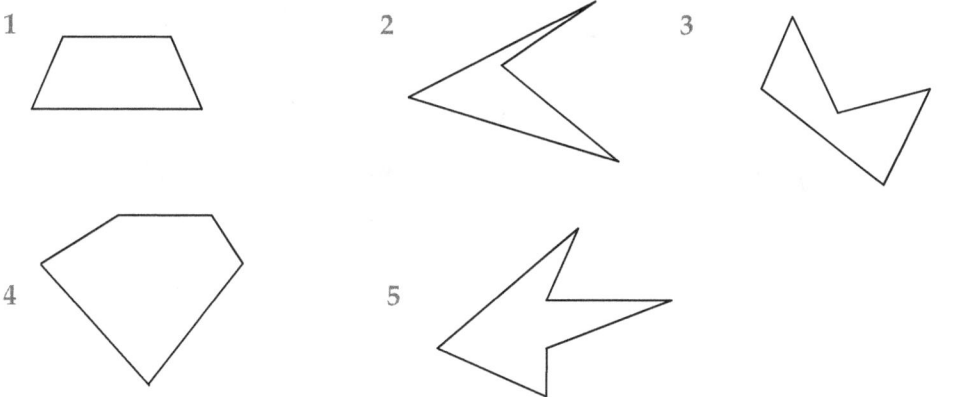

16.3.2 Regular and irregular polygons

Polygons with all sides and angles congruent are said to be regular, whereas those with varying sizes of angles and sides are said to be irregular polygons.

A square is a regular 4 sided figure, whereas a rectangle, rhombus, kite, trapezium, etc. are irregular four sided polygons.

Examples of regular polygons;

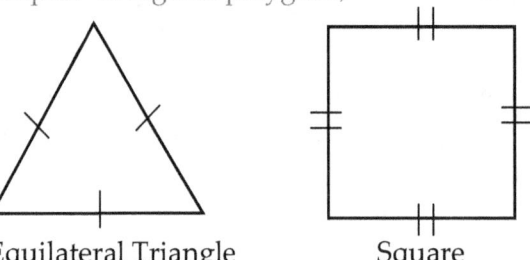

| Equilateral Triangle | Square | Regular Pentagon |

Examples of irregular polygons;

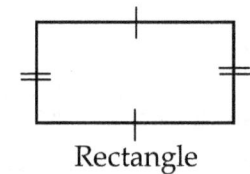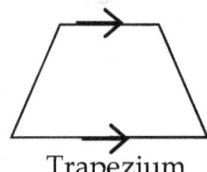

| Isosceles Triangle | Rectangle | Trapezium |

Exercise 16e

Given the figures below state which ones are polygons and state reasons why?

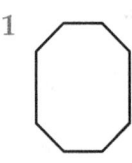

1   2   3   4   5

State whether each of the following polygons are regular or irregular.

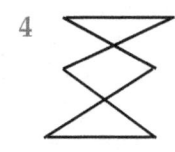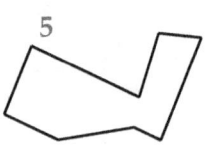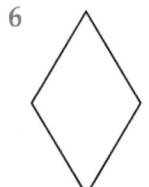

6   7   8   9   10

16.3.3 Triangles

A triangle is a three sided closed plane figure. The figure is a triangle with sides a, b, c and angles A, B, C.

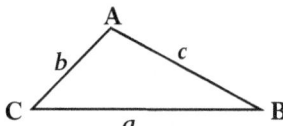

Types of Triangles

We basically have three types of triangles, that is, the *Equilateral*, *Isosceles* and *Scalene* triangles. However we can also categorize further an isosceles or scalene triangles to right angled triangles.

(a) *Equilateral triangle:* This is a triangle with all its sides equal and its angles congruent.

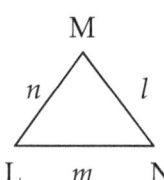

∢ M = ∢ L = ∢ N and sides $l = m = n$

For example, an equilateral triangle with sides = 3cm, and angles equal to 60° each.

(b) *Isosceles triangles:* An isosceles triangle is a triangle with only two of its sides equal and only two of its angles are congruent.

∢ P = ∢ R and length of side PQ equals to that of side QR.

(c) *Scalene triangle:* A scalene triangle is a triangle in which all its sides are different and all its angles are different.

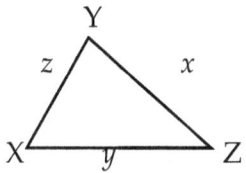

$x \neq y \neq z$

∢ X ≠ ∢ Y ≠ ∢ Z

(d) *Right angled triangles:* A right angled triangle is that triangle which has one of its angles equal to 90°. It can either be isosceles or scalene.

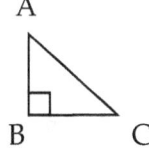
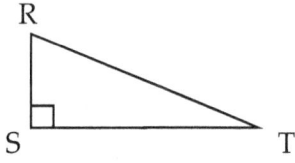

Triangle ABC is an Isosceles right angled triangle with the ∢ ABC equal to 90°, the sides AB = BC. RST is a scalene right angled triangle with the right angle ∢ RST.

16.3.4 Quadrilaterals

Quadrilaterals are four sided closed plane figures and they include squares, rectangles, rhombuses, kites, parallelograms, trapeziums, etc.

For quadrilaterals, the interior angle sum is equal to exterior angle sum which is equal to 360°. In the following figure, ∢ a, ∢ b, ∢ c and ∢ d are exterior angles and ∢ w, ∢ x, ∢ y and ∢ z are interior angles in the figure below.

∢ a + ∢ b + ∢ c + ∢ d = ∢ w + ∢ y + ∢ x + ∢ z.

Also;

∢ a + ∢ w = 180°,

∢ b + ∢ x = 180°,

∢ c + ∢ y = 180° and

∢ d + ∢ z = 180°

Square: A square is a regular quadrilateral with all sides equal and all angles congruent. The interior and exterior angles are equal to 90° each.
The diagonals of a square are equal and meet at 90°.

Rectangles: A rectangle has its interior and exterior angles equal to 90° each. In a rectangle the two opposite sides are equal and hence all squares are rectangles. Diagonals of the rectangle are equal.

Rhombus: A rhombus is a slanted square. All its sides are equal and the opposite interior angles are also equal.

Parallelogram: A parallelogram is a quadrilateral with two opposite sides equal and parallel and it is slanted. The opposite interior angles are equal.

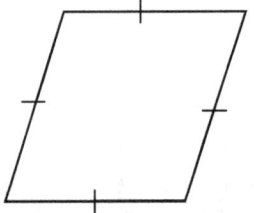

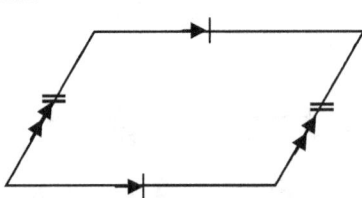

Kite: The Kite takes the form below;

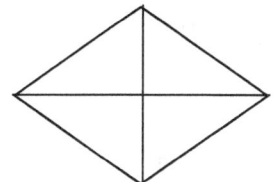

Trapezium: A Trapezium has two opposite sides unequal but parallel.

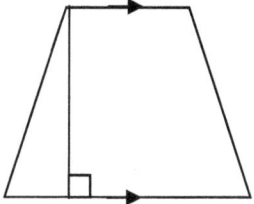
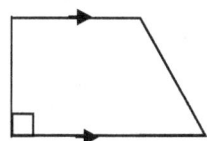

16.3.5 Polygons with more than four sides

Other polygons can be constructed with any number of sides. Below are some names for the figures with more than four sides.

No. of sides	Name
5	Pentagon
6	Hexagon
7	Heptagon
8	Octagon
9	Nonagon
10	Decagon
11	Nuo-decagon (Hendecagon)
12	Duo-decagon, etc.

All polygons have an exterior angle sum of 360°. Exterior angle size for a given regular polygon decrease with the increasing number of sides. For example, Given a regular Pentagon and Hexagon;

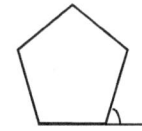
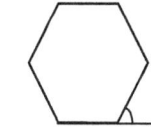

Regular Pentagon
Exterior angle is 72°

Regular Hexagon
Exterior angle is 60°

Exterior angles sizes

No. of sides	Name	Exterior angle
4	Square	90
5	Pentagon	72
6	Hexagon	60
7	Heptagon	51 $3/7$
8	Octagon	40

However interior angles for a given regular polygon increases with increasing number of sides, for example, given a regular pentagon and hexagon, we have;

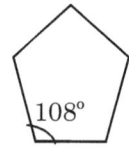

Regular Pentagon   Regular Hexagon

Interior angle sizes

No. of sides	Name	Interior angle
4	Square	90°
5	Pentagon	108°
6	Hexagon	120°
8	Octagon	135°

Note: The size of interior angle increases with the number of sides of a polygon.

Finding the interior and exterior angles of regular Polygons.

As you have seen the size of both Exterior or Interior angles depend on the number of sides a Polygon has.

(a) Exterior angles: We have already known that for every Polygon the exterior angle sum is 360°. For a regular polygon, to obtain the size of the exterior angle we divide the exterior angle sum (i.e. 360°) by the number of sides of a given polygon. If the exterior angle sum is S_e and the number of sides is n then the:

Exterior angle, $E = \dfrac{S_e}{n}$; but since $S_e = 360°$ and constant, we shall have;

$$\therefore \quad \sphericalangle E = \dfrac{360°}{n}$$

Example 1

Find the exterior angle size of a regular polygon with:

(i) 18 sides   (ii)   60 sides

Solution

(i) $\sphericalangle E = \dfrac{360°}{n}$

$\sphericalangle E = \dfrac{360°}{18°}$

$\sphericalangle E = 20°$

(ii) $\sphericalangle E = \dfrac{360°}{n}$

$\sphericalangle E = \dfrac{360°}{60°}$

$\sphericalangle E = 6°$

(b) Interior angles: We already know that a triangle has an interior angle sum of 180º. Now to know the interior angle sum of any polygon, we check the number of triangles we can obtain in a given polygon by drawing   diagonals to start at a particular vertex and spread to other vertices.

Remember that the diagonals should not cross each other.

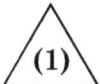 A triangle has no any diagonal.
This results to 1 triangle, angle sum 180°.

A rectangle has one diagonal.

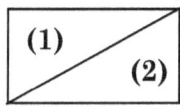 This results to 2 triangles, so interior angle sum is: (2 x 180°) = 360°.

A pentagon has 2 diagonals.

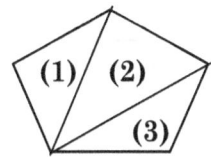 This results into 3 triangles so interior angle sum is: (3 x 180°)
= 540°

Activity 16.4

Find the interior angle sum of a polygon with eight sides, following the procedure shown below for six-sided figure: Since it has 4 triangles, then interior angle sum

= 4 x 180°

= 720°.

Therefore a six-sided figure has an interior angle sum of 720°.

To obtain the Interior angle for any polygon, you divide the interior angle sum by the number of sides the polygon has, for example, the interior angle of a hexagon is 720º, hence its interior angles will equal to;

$$= \frac{720°}{6°} \quad = 120°$$

Study the table next page carefully and complete it using the explanation earlier given.

Please remember to check yourself and learn the trend in the table.

No. of sides	No. of diagonals	No. of triangles	Interior angle sum	Interior angle
3	0	1	180°	60°
4	1	2	360°	90°
5	2	3	540°	108°
6	3	4		
7				
8				
9				
10				
15				
20				
25				
30	27			
35				
40				
45				
50				
100				
360				

The trend in finding the number of diagonals is that the number of sides exceeds the number of diagonals by 3.

So, we use the formula, $d = n - 3$; where d is number of diagonals, and n is number of sides.

Have you realized that an odd, n leads to an even d?

The trend in finding the number of triangles is that; the number of triangles exceeds the number of diagonals by 1.

Hence using the formula; $T = d + 1$ where, T is the number of diagonals.

From the table we can still obtain the relationship between the number of sides and interior angle sum.

Interior angle sum
$$= 180° \times T$$
$$= 180°T$$
but $T = d + 1$ and
$$d = n - 3;$$
$$T = [\,(n - 3) + 1\,]$$
$$= (n - 2)$$
So, interior angle sum:
$$= 180° \, T$$
$$= 180° \, (n - 2)$$
$$(\text{or } 90° \, (2n - 4).$$

Therefore, interior angle sum $= 90° \, (2n - 4)$.

Interior angle sum of pentagon is
$$= 90° \, (2 \times 5 - 4)$$
$$= 90° \, (6)$$
$$= 90° \times 6$$
$$= 540°;$$

Interior angle sum of hexagon is
$$= 90° \, (2 \times 6 - 4)$$
$$= 90° \times 8$$
$$= 720°; \text{ etc.}$$

Exercise 16f

1 Given the figure below, name all the regular polygons that can be obtained from it.

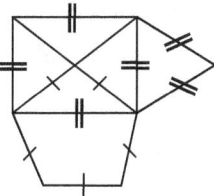

Find the interior angle sum of polygons with:

2   8 sides   3   11 sides

4   12 sides   5   24 sides

6 If the exterior angle of a regular polygon is 24°.

  (i) How many sides has the polygon?

  (ii) What is its interior angle sum?

16.4 Polyhedrons

Polyhedrons are three dimensional figures formed by union of polygons. The polygons which form a polyhedron must be simple, intersect at edges and the figure (polyhedron) must be closed.

Cube: A cube is a solid figure obtained from a combination of 6 squares meeting at edges and vertices.

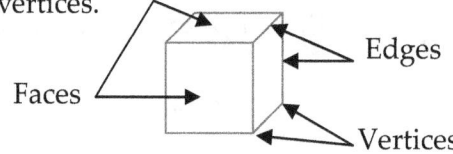

Faces

Edges

Vertices

Cuboid: A cuboid is a solid figure obtained from a combination of 6 rectangles meeting at edges and vertices.

Remember that, all squares are rectangles. Likewise, all cubes are cuboids.

Prisms: Prisms are polyhedrons with a pair of opposite faces which are parallel. Prisms are classified by the shape of the polygons that form the parallel faces. For example, triangular quadrangular, pentagonal, hexagonal and octagonal prisms, correspond to prisms formed by polygons with 3, 4, 5, 6 and 8 sides respectively, forming their cross sectional faces.

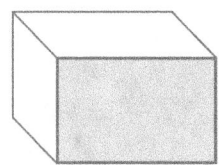 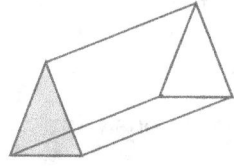 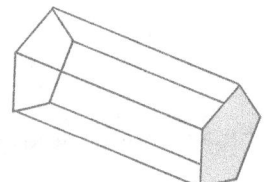

Quadrangular prism   Triangular prism   Pentagonal prism

The side faces are called lateral faces. In a case where the lateral faces are perpendicular to the base, the prism is known as a right prism and where the lateral face form an angle to the base we have oblique prisms.

Regular polyhedron: Regular polyhedrons are those in which all faces and polyhedral angles are congruent. Polyhedral angles are formed by at least three faces and when three regular triangles meet at a vertex they form a polyhedral angle of 180°. A polyhedral angle is that which is formed at the vertex when planes/faces forming a polyhedron meet there.

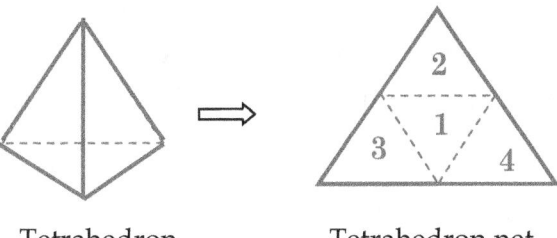

Tetrahedron   Tetrahedron net

A tetrahedron is formed from 4 equilateral triangular faces. At the vertex, there are 3 angles of the triangle, each equaling to 60° meeting, hence forming a polyhedral angle of 180°.

If this net is cut out of paper and folded along the dotted lines, a solid figure called a tetrahedron is formed as shown in the figure to the left.

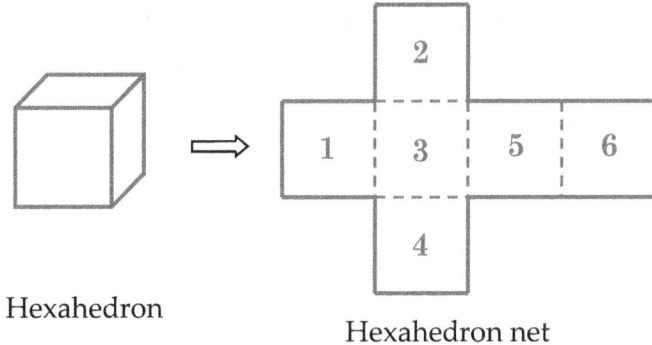

Hexahedron

Hexahedron net

A hexahedron is formed from 6 square faces. At the vertex, there are 3 angles of the square, each equaling to 90º meeting, hence forming a polyhedral angle of 270°.

If this net is cut out of paper and folded along the dotted lines, a solid figure called a hexahedron is formed as shown in the figure to the left. A hexahedron is also known as the *Cube*.

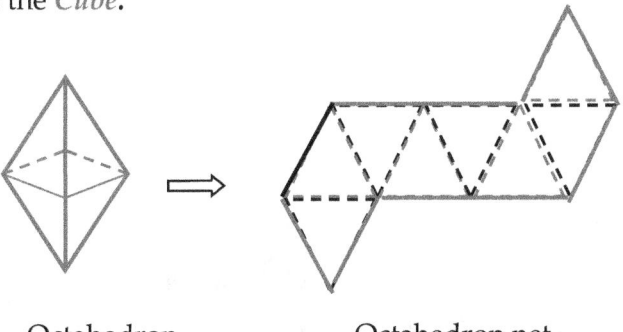

Octahedron   Octahedron net

An Octahedron is formed from 8 equilateral triangle faces. At the vertex, there are 4 angles of the triangle, each equaling to 60° meeting, hence forming a polyhedral angle of 240°.

360° is the largest polyhedral angle we can have.

A net of a polyhedron is also known as the *development*, e.g. the development of an Octahedron (for Octahedron net).

16.5 Other Polyhedrons

Pyramids: A pyramid is a 3 dimensional figure which has a polygon for the base and triangular faces that meet at a point (vertex). It has a line from the vertex called altitude (or height) which is perpendicular to the base.

Pyramids are classified by the shape of the base, for example; triangular, rectangular, pentagonal or hexagonal pyramids, etc.

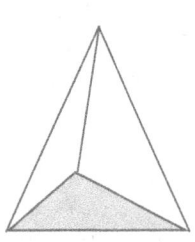   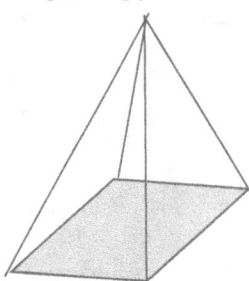   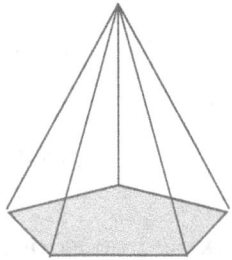

Triangular Pyramid   *Square based pyramid*   *Pentagonal based pyramid*

Cylinders: These have curved surfaces and circular bases.

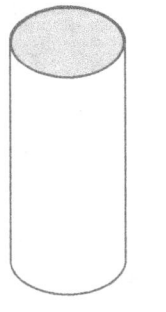   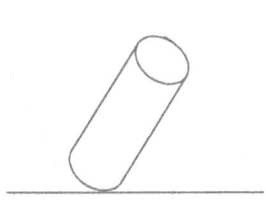

Oblique cylinder

Right cylinder

Sphere   *Hemisphere*   *Cone*

 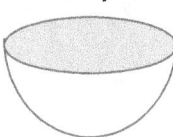 

A sphere is a round solid figure with equal distances from its centre to any part of its surface. A hemisphere is half of the sphere.

A cone is a solid figure the has a circle at the base and a surface that comes to a point (the vertex).

Exercise 16g

1 Draw the developments of the tetrahedron, hexahedron and octahedron.

   (a) Cut them out.   (b) Fold them to obtain the polyhedron (use glue)

2 Draw the development of the icosahedron shown below:

Development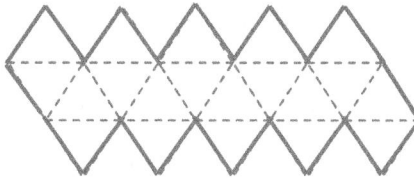

20 Equilateral triangles

Trace and cut out this development and fold it to form the icosahedron.

REVISION EXERCISE 16

1   (a) State the types of the following angles:

   (i) 90°   (ii) 38°   (iii) 123°   (iv) 352°

   (b) State the relationship between each pair the given angles.

   (i) 39° and 46°   (ii) 137° and 43°

   (iii) 46° and 46°   (iv) 45° and 45°

   (c) Given angle PQR, where ∢ PQX + ∢ RQX = ∢ PQR.

What is the relationship between ∢ RQX and ∢ PQX?

2   Find the values of angles represented by letters in degrees.

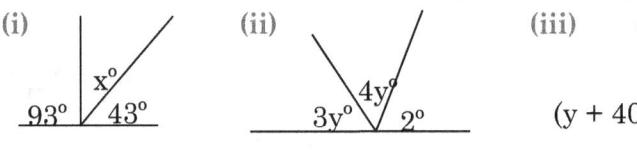

(i)   (ii)   (iii)

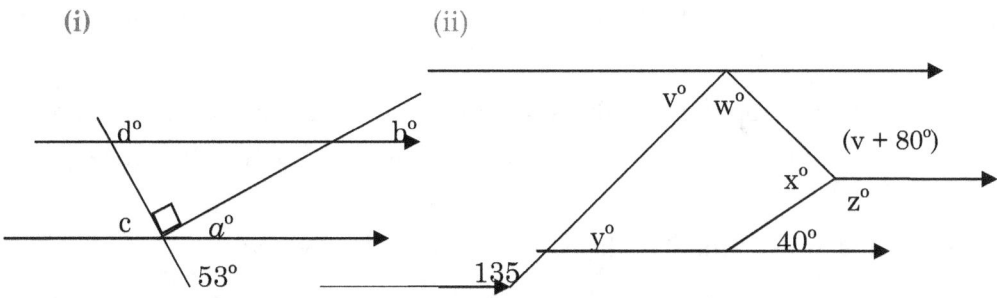

3   Find the sizes of angles represented by letters in the following figures and give reasons.

(i)   (ii)

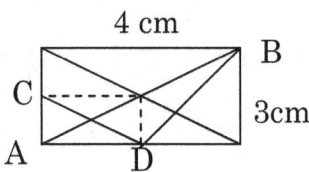

4   Find the values of lengths of the lines in the figures below:

(i) Lines AB, CD and BD   (ii)   Given that figure ABCDE is a regular pentagon, find; AD, AB and BX

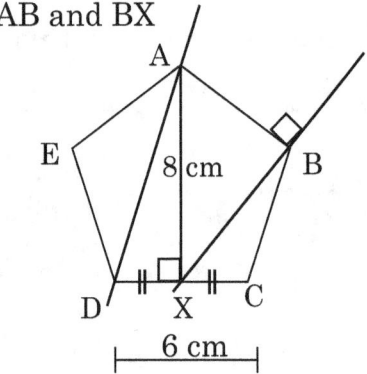

5   If the interior angle of a regular polygon is less than twice its exterior angle by 36°.

Find: (i) Its exterior angle.   (ii) Its interior angle.

(iii) Its interior angle sum.

GRAPHS AND EQUATIONS OF LINES

INTRODUCTION

Lines on a grid have different properties such as the length (or distance) of the line, the points lying on the line and the description of the line. The line distance can be in the x, y or both directions. A line on a grid can be described by its equation, hence the equation of line.

A line can be described according to the points lying on it and the same points can used to draw it on a grid.

17.1 Distance on grid

Consider the grid below for the following;

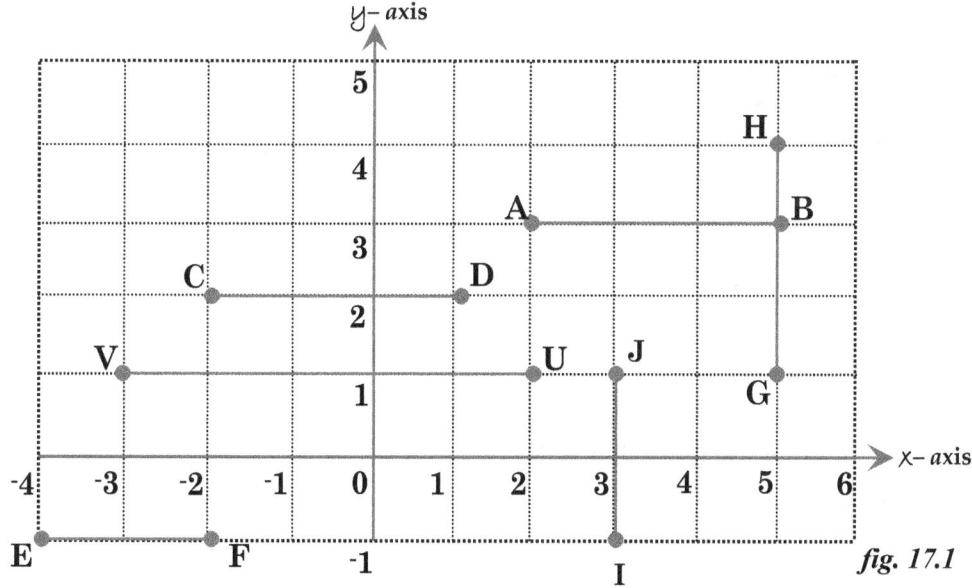

fig. 17.1

17.1.1 Distance between x-coordinates

Consider lines AB, CD and EF. You can find the distance of these lines by counting the units they cover from the grid. So, AB = 3 units, CD = 3 units and EF = 2 units.

However if we donJt have the grid we can use the coordinates of the points that form the line to find the lengths of these lines.

For line AB, the coordinates are A(2, 3) and B(5, 3). When we look at the x-coordinate the change is 3 (that is, 5 – 2) and there is no change in the y- coordinates (that is, 4 – 4 = 0). So, the distance of the line AB is 3 units.

Line CD has coordinates; C(-2, 2) and D(1, 2), we therefore have:
Change in x-coordinates as (1 – -2) = 3.
Change in y is (2 – 2) = 0 (no change in y-coordinate). Therefore the length CD is 3 units.

Line EF has coordinates; E(-4, -1) and F(-2, -1), we therefore have:
Change in x is (-1 – -4) = 2.
Change in y is (-1 – -1) = 0 (no change in y-coordinates). Therefore the length CD is 2 units.

Activity 17.1

Find the distances of the lines formed by the following points (Explain your answers):
1 Line PQ with coordinates, P(6, 4) and Q(-6, 4)
2 Line RS with coordinates, R(-2, 4) and S(-2, 4)

Exercise 17a

1 On a grid plot the following points:
   (i) A (2, 1)   (ii) B(-3, 4)   (iii) C(-8, 2)   (iv) D(5, -6)   (v) E(3, -4)
   (vi) F(-2, -4)   (vii) G(2, -6)   (viii) H(3, 4)   (ix) I(4, 1)   (x) J(-3, 2).

2 On the grid join the points to obtain the lines,
   (i) AI   (ii) BH   (iii) CJ   (iv) DG   (v) EF

3 By counting the units on the grid find the lengths of the lines in question 2 above .

4 By use of calculations find the lengths of the lines in question 2 above.

17.1.2 Distance between y-coordinates

Consider lines GH and IJ on the grid in fig. 17.1, you can find the lengths of these lines by counting the units from the grid. So GH = 3 units and IJ = 2 units.

However, if we donJt have the grid we can use the coordinates of the points that form these lines to find the lengths of these lines. For line GH, the coordinates are G(5, 1) and H(5, 4). There is no change in the x- coordinates (that is, 5 – 5 = 0), and the change in the y-coordinates is 3 (that is, 4 – 1 = 3). Therefore, the length GH is 3 units.

For line IJ, I(3, -1), J(3, 1). No change in x- coordinates.
Change in y- coordinates is (1 – ˉ1 = 2). Therefore length of IJ = 2 units.

Exercise 17b

1   On a grid plot the following points:
   (i) K(0, 2)   (ii) L(ˉ2, ˉ3)   (iii) M(3, ˉ5)   (iv) N(7, ˉ2)   (v) P(3, 4)
   (vi) Q(7, ˉ3)   (vii) R(7, 2)   (viii) S(3, ˉ1)   (ix) T(ˉ2, 0)   (x) U(0, 5)

2 On the grid join the points above in 1 to obtain the following lines:
   (i) UK   (ii) LT   (iii) MS   (iv) NR   (v) QN

3 By counting units find the lengths of these lines above in question 2.

4 By use of calculation, find the length of these lines in question 2 above.

17.1.3 Distance as a scalar

Given the points U(2, 1) and V(ˉ3, 1), we can see that there is no change in y-coordinates and the change in x-coordinates is (ˉ3 – 2) = -5. Plotting these points on the grid as shown in the fig. 17.1, we can count the units of length for the line UV to obtain 5.

What do you realize? You will find that when you calculate, you get 5, and when you count you get ˉ5.

The answer -5 means that the counting of units from U to V is 5 units towards the left hand side of the x-axis.

Distance is a scalar quantity, therefore the direction sign is not necessary since distance in any direction is equal as long as it is of the same magnitude. In this case, the distance of UV is 5 units.

17.1.4 Distance formula

When you look at lines KL and MN on the grid in fig.17.2 that follows, it is difficult to read the length directly from both the x and y axes on the grid.

So, in this case we resort to the use of a formula to find the distance of a given line with given coordinates on the grid.

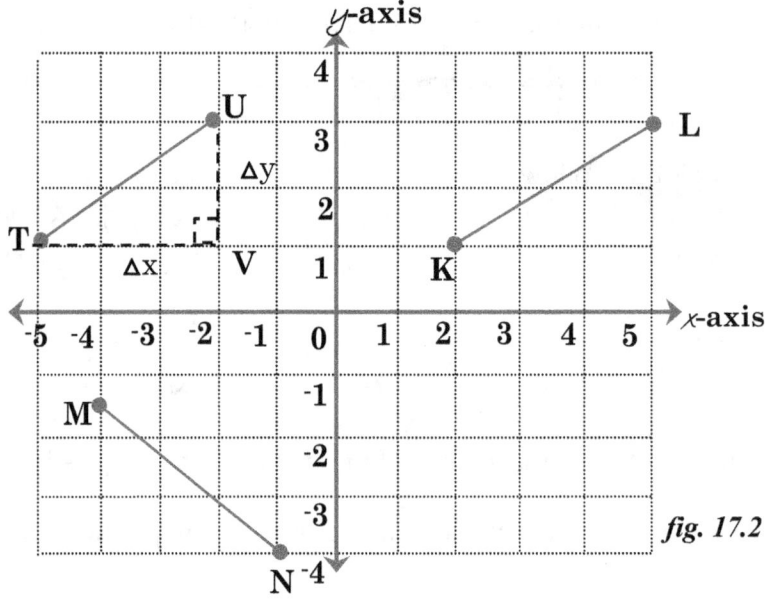

fig. 17.2

Let the change in x-coordinates be Δx and the change in y-coordinates be Δy.

If we have two points:

$T(x_1, y_1)$ and $U(x_2, y_2)$, we can use a formula to find the length of the line TU.

To understand how the formula is obtained, study the triangle TUV which is derived from the line TU as shown on the diagram in fig. 17.2.

Let us extract the triangle TUV from the grid and use it as follows:

Triangle TUV:–

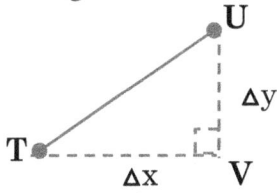

From this figure, triangle TUV is a right angled triangle.

Therefore the length TU is given by:

$TU = \sqrt{VT^2 + UV^2}$.

(As it originates form $TU^2 = TV^2 + VU^2$)

But $TV = \Delta x$ and $VU = \Delta y$. And also, $\Delta x = (x_2 - x_1)$ and $\Delta y = (y_2 - y_1)$

Generally, the length, d, for line TU is:

$d = \sqrt{VT^2 + UV^2}$

$d = \sqrt{(\Delta x)^2 + (\Delta y)^2}$

$d = \sqrt{(x_2 - x_1)^2 + (y_2 - y_1)^2}$

For line TU;

$x_1 = -5$, $x_2 = -2$, $y_1 = 1$ and $y_2 = 3$

$TU = \sqrt{(-2 - -5)^2 + (3 - 1)^2}$

$TU = \sqrt{(3)^2 + (2)^2}$

$TU = \sqrt{9 + 4}$

$TU = \sqrt{13}$ (or 3.6056) units.

For the lengths of lines KL and MN using the same formula, we can have the working as follows;

KL; $x_1 = 2$, $x_2 = 5$,
    $y_1 = 1$ and $y_2 = 3$

$KL = \sqrt{(5 - 2)^2 + (3 - 1)^2}$

$KL = \sqrt{(3)^2 + (2)^2}$

$KL = \sqrt{9 + 4}$

$KL = \sqrt{13}$ (or 3.6056) units.

MN; $x_1 = -4$, $x_2 = -1$,
    $y_1 = -1.5$ and $y_2 = -4$

$MN = \sqrt{(-1 - -4)^2 + (-4 - -1.5)^2}$

$MN = \sqrt{(3)^2 + (-2.5)^2}$

$MN = \sqrt{9 + 6.25}$

$MN = \sqrt{15.25}$ (or 3.9051) units.

Exercise 17c

Given the pairs of the points below, find the distance of the line joining each of the two points in the pair.

1   A(5, -4) and B(8, -1)   2 C(1, 1) and D(2, -3)   3 E(0, 3) and F(-1, 2)

4   G(-4, 11) and H(-2, 2)   5 I(4, 7) and F(-1, 2)   6 K(-6, 1) and L(-11, 2)

7   M(5, -2) and N(-7, 4)   8 P(-11, 3) and Q(-4, 11)   9 R(-3, 2) and S(8, -2)

10 T(-4, 0) and U(7, 3)

17.2 Equations of lines

Carefully study the lines on the grid below:

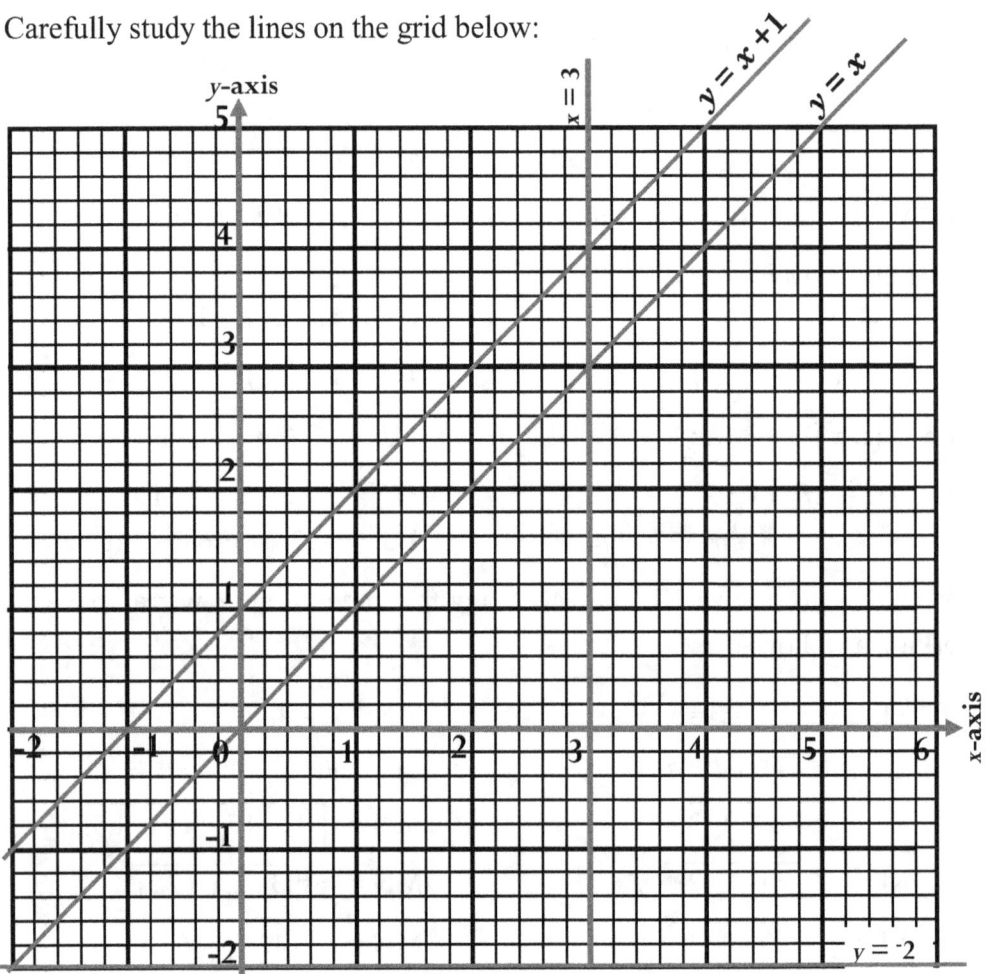

Tabulation:

We can tabulate the coordinates of the points on each of the lines; $x = 3$, $y = -2$, $y = x$ and $y = x + 1$ for the values of x ranging from -4 to $+5$, (or $-4 \leq x \leq +5$).

In each case substitute the value of x as given into the equation of line to obtain its corresponding y-value.

Study the values tabulated for the plotted and drawn lines as follows.

For $x = 3$;

We realize that all the x-coordinates are equal to 3 and the y-coordinates vary.

x	3	3	3	3	3	3	3	3	3	3
y	-4	-3	-2	-1	0	1	2	3	4	5

The coordinates can be listed as (3, ⁻4), (3, ⁻3), (3, ⁻2), etc.

In this case where $x = 3$, as a constant value of x, the y-values change continuously.

For $y = ⁻2$;

Here we realize that all the y-coordinates are equal to ⁻2. In this case where $y = ⁻2$, as a constant value of y, the x-values change continuously.

x	⁻4	⁻3	⁻2	⁻1	0	1	2	3	4	5
y	⁻2	⁻2	⁻2	⁻2	⁻2	⁻2	⁻2	⁻2	⁻2	⁻2

For $y = x + 1$

x	⁻4	⁻3	⁻2	⁻1	0	1	2	3	4	5
y	⁻3	⁻2	⁻1	0	1	2	3	4	5	6

From $x + 1 = y$

$⁻2 + 1 = ⁻1$
$⁻4 + 1 = ⁻3$ $⁻1 + 1 = 0$
$⁻3 + 1 = ⁻2,$ $0 + 1 = 1,$ etc.

Exercise 17d

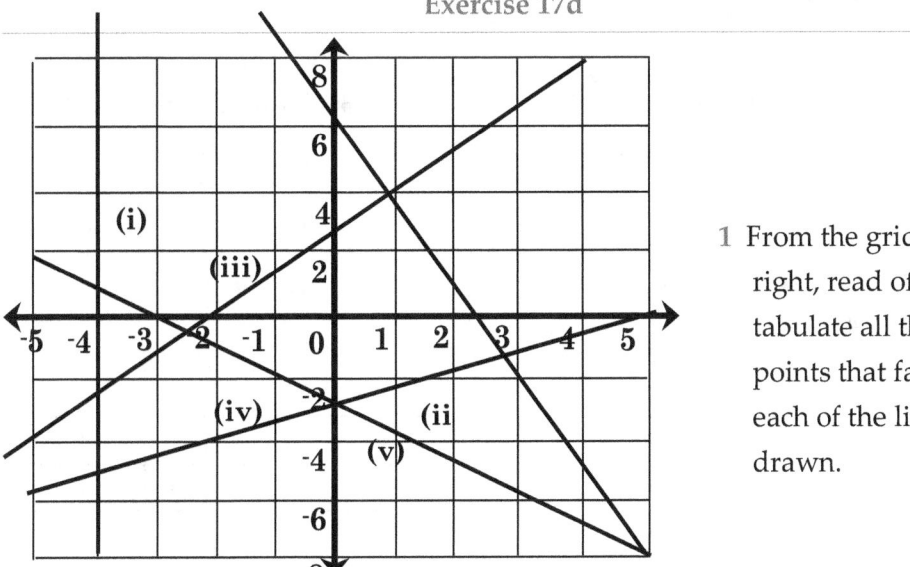

1 From the grid to right, read off and tabulate all the points that fall on each of the lines drawn.

2 In each of the tables you have made for the lines (i), (ii), (iii), (iv) and (v) add four more points on the positive end for each of them.

17.2.1 Determining points which belong to a line

For a line, $y = x + 3$, we can obtain the value of y if we know the value of x. For the coordinates of the points on $y = x + 3$, we choose any value of the x-coordinate, then substitute them in the equation of the line to obtain the corresponding y-coordinate.

Let us choose the following values of x for $y = x + 3$;

$x = \{ \text{-}4, \text{-}3, \text{-}2, \text{-}1, 0, 1, 2, 3, 4\}$

Now substituting for x: $\quad x + 3 = y$

$\text{-}4 + 3 = -1;$	$-1 + 3 = 2;$	$2 + 3 = 5;$
$\text{-}3 + 3 = 0;$	$0 + 3 = 3;$	$3 + 3 = 6;$
$\text{-}2 + 3 = 4;$	$1 + 3 = 4;$	$4 + 3 = 7.$

Now we can tabulate as follows:

x	-4	-3	-2	-1	0	1	2	3	4
y	-1	0	1	2	3	4	5	6	7

Remember that a point will always belong to a set of points for a particular line only and only if it certifies the equation of that line.

Example 1

Given the x-coordinates such that $x = \{\text{-}2, \text{-}1, 0, 1, 2, 3, 4\}$, find the y-coordinates of the following lines: (i) $y = 2x$ (ii) $y = 3x - 2$ (iii) $2y = x + 3$

Solution

(i) For line $2x = y$

$2(\text{-}2) = \text{-}4 \qquad 2(1) = 2$
$2(\text{-}1) = \text{-}2 \qquad 2(2) = 4$
$2(0) = 0 \qquad 2(3) = 6$
$\qquad\qquad\quad 2(4) = 8,$

in the table $y = 2x$

x	-2	-1	0	1	2	3	4
y	-4	-2	0	2	4	6	8

(ii) For line $3x - 2 = y$

$3(\text{-}2) - 2 = (\text{-}6 - 2) = \text{-}8 \qquad\qquad 3(0) - 2 = (0 - 2) = \text{-}2$
$3(\text{-}1) - 2 = (\text{-}3 - 2) = \text{-}5 \qquad\qquad 3(1) - 2 = (3 - 2) = 1$

$3(2) - 2 = (6 - 2) = 4$

$3(3) - 2 = (9 - 2) = 7$

$3(4) - 2 = (12 - 2) = 10$, in table $y = 3x - 2$

x	-2	-1	0	1	2	3	4
y	-8	-5	-2	1	4	7	10

(iii) For line $2y = x + 3$

Here first make y the subject

$$\frac{2y}{2} = \frac{x+3}{2} \Rightarrow y = \frac{x+3}{2}$$

$$\frac{x+3}{2} = y$$

$x = -2;$ $\dfrac{-2+3}{2} = \dfrac{1}{2}$

$x = -1;$ $\dfrac{-1+3}{2} = 1$

$x = 0;$ $\dfrac{0+3}{2} = \dfrac{3}{2} = 1\frac{1}{2}$

$x = 1;$ $\dfrac{1+3}{2} = 2$

$x = 2;$ $\dfrac{2+3}{2} = \dfrac{5}{2} = 2\frac{1}{2}$

$x = 3;$ $\dfrac{3+3}{2} = \dfrac{6}{2} = 3$

$x = 4;$ $\dfrac{4+3}{2} = \dfrac{7}{2} = 3\frac{1}{2}$

Table for line, $2y = x + 3$

x	-2	-1	0	1	2	3	4
y	½	1	1½	2	2½	3	3½

We can now use these coordinates to plot the line $2y = x + 3$.

Always remember that, if we know the y-coordinate we can still find the x-coordinate; for $2y - 3 = x$, where $y = 4$,

We have: $2(4) - 3 = x$

$8 - 3 = x$

$5 = x$

So, we have the point (5, 4) on the line $2y = x + 3$.

Exercise 17e

Given the x-coordinates ranging from $x = -5$ to $x = 5$, tabulate the coordinates of the following lines:

1 $y = x + 2$ 2 $x + y = 3$ 3 $2x - y = 2$ 4 $y = 6 + x$

5 $3y = x - 1$ 6 $2x - y = -1$ 7 $\dfrac{x+2}{3} = y$ 8 $\dfrac{8-x}{y} = 2$

On a grid draw and name the lines:

9 $x + y = 5$ 10 $y = 5x - 2$; for x such that $-3 \le x \le 4$.

17.2.2 Obtaining the equation of line from points on the line

Writing the equation of a line given coordinates of points on that line consider the table below.

x	1	2	3	4	5
y	4	8	12	16	20

When we study carefully the coordinates in the table you will see some relationship between the x and y coordinates.

The coordinates for the points are; (1, 4), (2, 8), (3, 12), (4, 16) and (5, 20). Have you realized that the y-value is 4 times the x-value? That is, $y = 4x$.

Check the equation by substituting the coordinate values in the equation.

Example 1

Given the coordinates in the tables below, find the equations for the lines on which they fall:

(i)

x	1	2	3	4
y	4	8	12	16

(ii)

x	1	2	3	4
y	·1	1	3	5

(iii)

x	1	2	3	4
y	4	7	10	13

Solution

Here our task is to look for the relationship between the x-coordinate and the y-coordinate for every particular point.

So we need to remember the knowledge of number sequences.

(i)

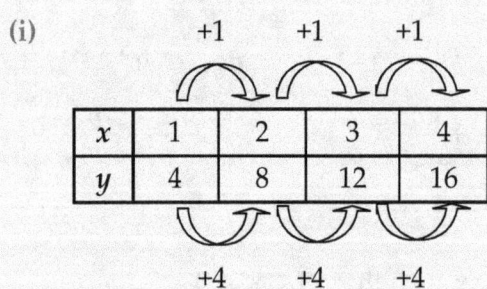

When you realize that the common difference in x-value is 1, then multiply x by 4, the common difference in y-value, then check .

4x	4	8	12	16
y	4	8	12	16

Now the results for 4x equals to y, so we have the equation: $y = 4x$.

(ii)

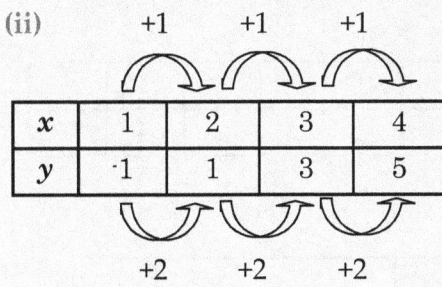

Since the common difference in x-values is 1, multiply x by the common difference of y-values, that is,

2x	2	4	6	8
y	-1	1	3	5

Are the 2x values equal to the corresponding y-values? No, so now look for the relationship between 2x and y. Since the common difference between 2x and y values is the same as 2.

Have you realized that the 2x value exceeds its corresponding y-value by 3 in each case?

Now check by subtracting 3 from the 2x value to obtain the corresponding y-value in each case.

2x	2	4	6	8
-3	-3	-3	-3	-3
y	-1	1	3	5

Now, we realize that from this table 2x and y values correspond with the x and y values in the given table with a relationship.

The equation of the line therefore is $y = 2x - 3$.

(iii)

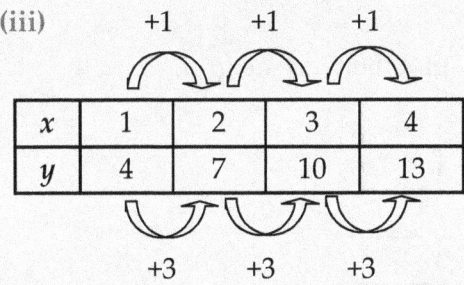

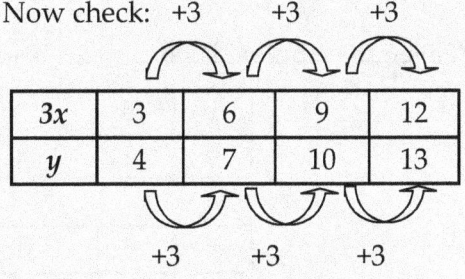

What is the relationship between 3x and y? we add 1 to 3x to obtain y. Remember that the moment the common difference is the same in the two values being compared then they have a relationship. The equation of the line therefore is $y = 3x + 1$.

Exercise 17f

Given the following coordinates of the lines find the equations of these lines.

1

x	1	2	3	4
y	3	2	1	0

4

x	0	1	2	3	4
y	5	2	-1	-4	-7

2

x	1	2	3	4	5
y	4	3	2	1	0

5

x	-1	-4	-7	-10
y	1	2	3	4

3

x	1	2	3	4
y	1	3	5	7

6

x	-4	-3	-2	-1
y	-5	-3	-1	1

REVISION EXERCISE 17

1 Given the points, A(3, 1), B(6, 2), C(-1, 0), D(-1, 3) and E (-¼, 2)

   (*a*) Plot them on a grid.

   (b) Find lengths of the lines: (i) AB   (ii) CD   (iii) BE

                                       (iv) AC   (v) CE

2 Given the tabulated coordinates below, plot them on the grid.

   (*a*) $y = 2x + 1$

x	0	1	2	3
y	1	3	5	7

   (b) $y = x - 1$

x	0	1	2	3
y	-1	0	1	2

3 Tabulate the coordinates of the following lines given on the grid next page.

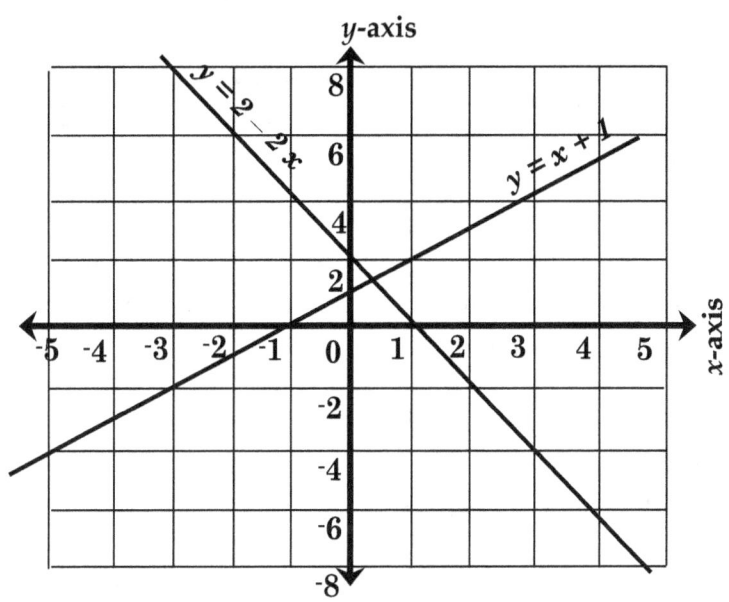

4 Tabulate eight pairs of coordinates for each of the following lines:

   (i) $y = 2x - 3$ (ii) $y = 8 - 5x$ (iii) $2y = x - 1$

5 Given the tabulated coordinates, find the equations of lines on which these coordinates lie.

(i)

x	-1	0	1	2
y	-1	0	1	2

(iii)

x	1	2	3	4
y	3	4	5	6

(ii)

x	-2	-1	0	1
y	-2	-2	0	2

(iv)

x	-1	0	1	2	3
y	-5	-3	-1	1	3

18 SYMMETRY, REFLECTION AND CONGRUENCE

INTRODUCTION

If you view yourself in a mirror and raise your right arm, your image in the mirror will have raised its right arm! Have you ever realized this? Such concepts have always helped us in daily lives.

Ambulances are always labeled " ƎƆИA⅃UBMA ", and this helps to driver ahead the ambulance to correctly see and read the word as viewed in his driving mirror such that he or she gives way.

When a stamp is being made the words on it are reversed the same as shown for the ambulance. In this arrangement we can clearly stamp documents.

Stamp Engraving

Stamp on document

18.1 Symmetry in two dimensions

Given the figures below, can you notice the common features that exist in each? Look carefully at the figures given in fig. 9.1 separately.

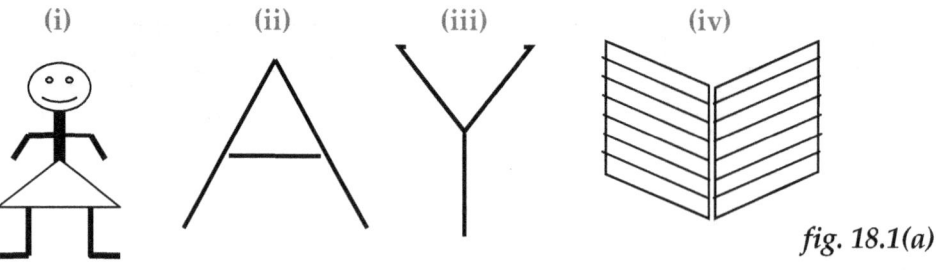

(i)   (ii)   (iii)   (iv)

fig. 18.1(a)

In fig. 18.1, we realize that the right hand side in each of the parts (i), (ii), (iii) and (iv) has the same shape and size as the left hand side. To separate the two sides we may use the dotted line in the middle as shown next page:

Left ⋮ Right

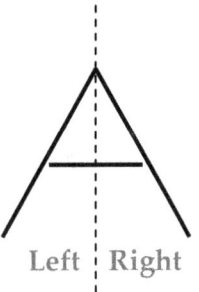

Left ⋮ Right

fig. 18.1(b)

Activity 18.1

1 Draw lines separating the figures (iii) and (iv), in fig. 18.1(a) into the similar parts of the same shape and size.

2 Think of 10 figures which have the right and the left hand sides with the same shape and size, draw them and the lines that separate each of them into two parts of the same shape and size.

We realize that in fig. 18.2(*a*), the figures have the top and bottom parts of the same shape and size.

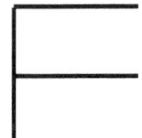
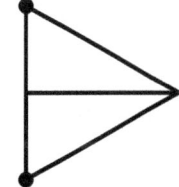

fig. 18.2(a)

Top

Bottom

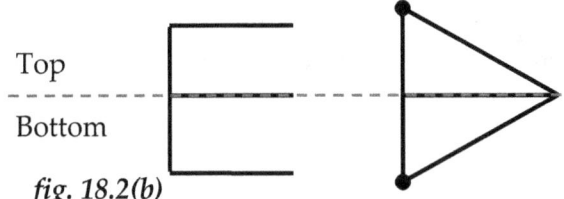

We can draw dotted lines in the middle to separate the two parts as shown in fig. 18.2(*b*).

fig. 18.2(b)

Activity 18.2

Think of 5 figures which have the top and bottom parts of the same shape and size. Draw them and the lines that separate them into two equal parts of the same shape and size

Can we have a shape divided into more than two equal parts?

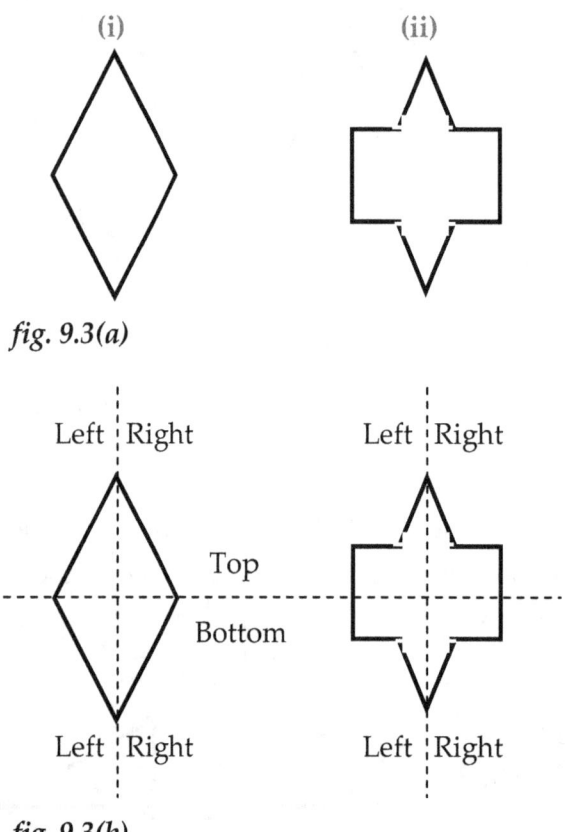

(i)

(ii)

fig. 9.3(a)

Left ¦ Right   Left ¦Right

Top

Bottom

Left ¦ Right   Left ¦Right

fig. 9.3(b)

In fig. 18.3(a), we realize that in these figures the top and bottom parts are similar in shape and size. And the left and right sides are the same in shape and size.

We now have two lines which can divide each figure into two equal parts of the same size and shape, see fig. 9.3(b).

Each figure results into 4 similar parts in shape and size as shaded.

Activity 18.3

Think about 3 figures, which have two lines dividing them into equal parts of the same shape and size. Draw them and the lines diving them into two equal parts of the same shape and size.

How many lines can divide the following figures into equal parts of the same shape and size?

For the letter (F) we have no line that can divide it into equal parts of the same shape and size. Whereas for the circle the number of lines that can divide it into equal parts is inexhaustible. All diameters taken at any angle will divide the circle into two equal parts.

Exercise 18a

Given the figures below draw dotted lines in each case to separate each figure into two equal parts in shape and size where possible

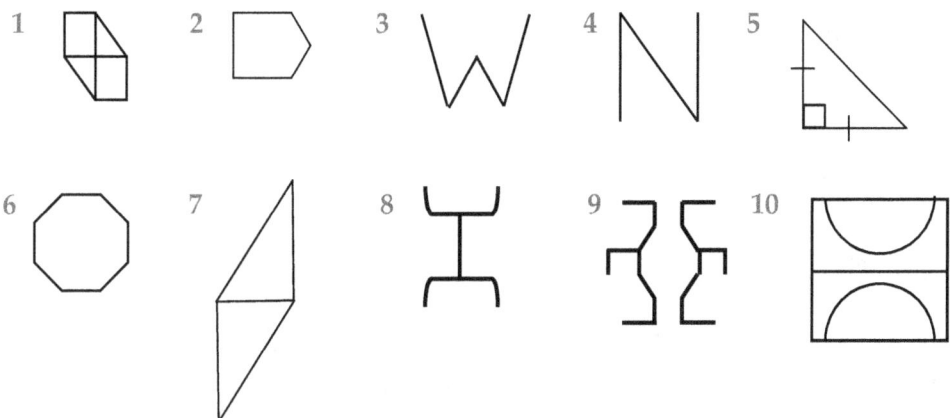

18.2 Folding symmetry

Trace the following shapes and cut them out. Try to fold them such that one part fits exactly on the other.

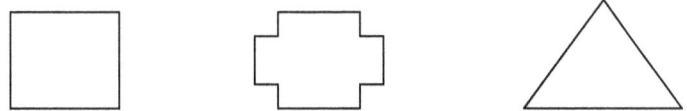

Each time you fold about a certain line and the part fits on the other exactly. Draw there a line. We shall then have the following:

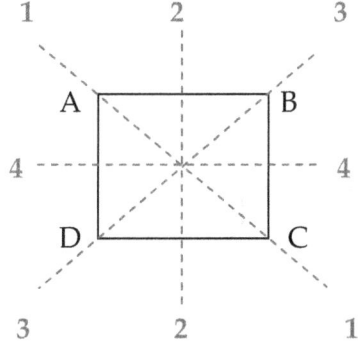

There are 4 ways in which we can fold a square ABCD such that one part fits exactly on the other; about the lines 1-1, 2-2, 3-3 and 4-4.

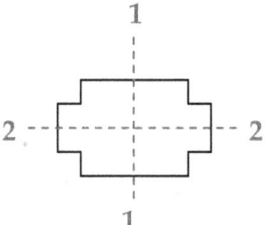

There are two ways in which one part of this figure can exactly fit onto the other. Make sure you are clear with how this happens

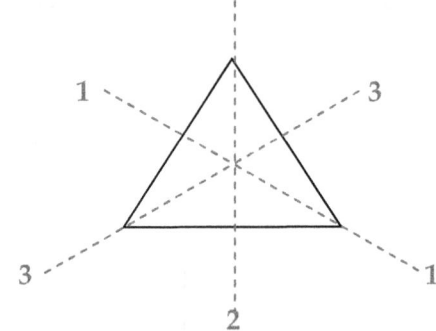

There are three ways in which one part of this figure can fit exactly on the other, if folded about certain lines as shown above.

Activity 18.4

Think of 2 shapes, draw them, cut them out and try to fold them such that a part fits exactly on the other. List the ways each part fits exactly on the other.

Exercise 18b

Trace out the following shapes, cut them out and fold each of them to obtain a flap fitting exactly on the other. In each case find how many ways does each flap fit on the other exactly of the same shape and size.

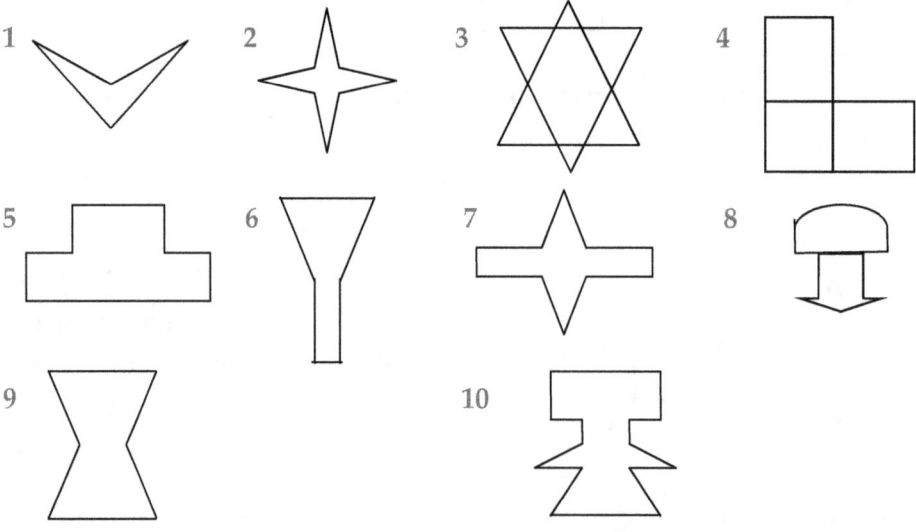

18.3 Axes of symmetry

In symmetry we have already realized that when we divide a shape into two similar parts of the same shape and size, we do it along a line. And when we fold a shape drawn on a piece of paper in order to have two parts which exactly fit on each other, we also do it along a line. In both cases, the line divides the figure into two similar parts.

An axis of symmetry therefore, is that line which divides a figure into two similar parts of the same shape and size. It is also called a *line of symmetry*.

Given the figures (i), (ii), (iii), (iv) and (v) we can tell the number of lines of symmetry each has.

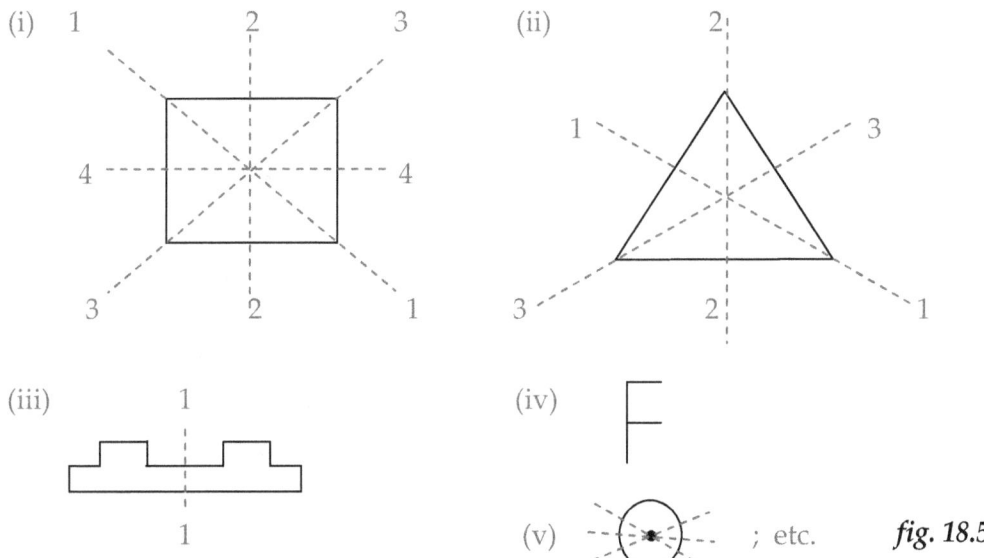

fig. 18.5

From the figure 18.5, we have the figures;

        (i) has 4 lines of symmetry.

        (ii) has 3 lines of symmetry.

        (iii) has 1 line of symmetry.

        (iv) has no line of symmetry.

        (v) has an infinitive number of lines of symmetry.

In this case when ever you cut through the centre to form a diameter you obtain a line of symmetry.

State lines of symmetry each figure has in the following:

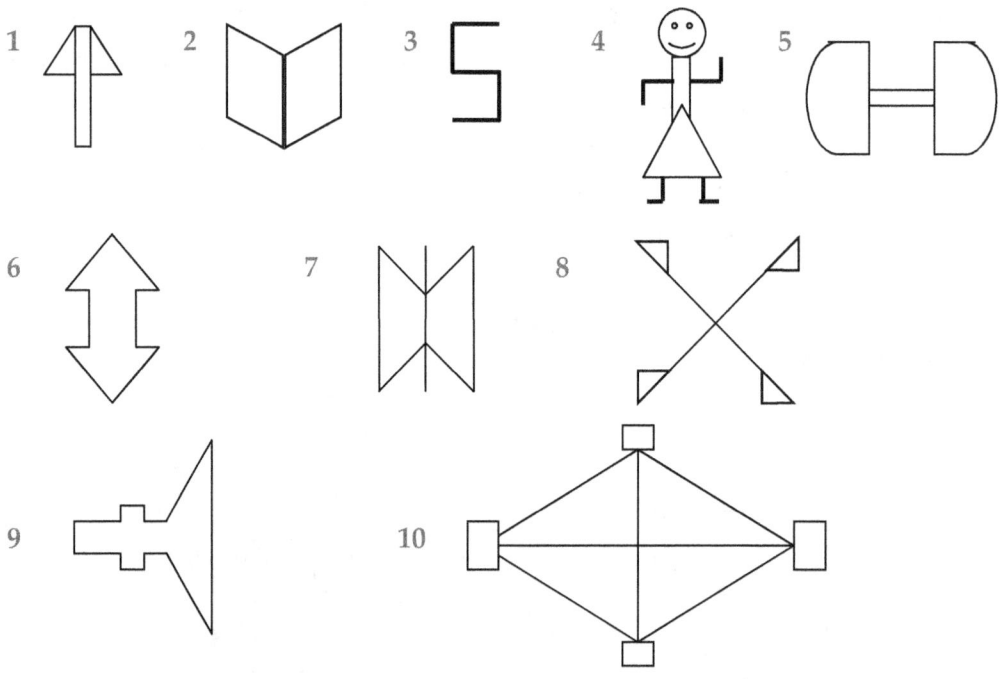

Study the figure 18.6 below.

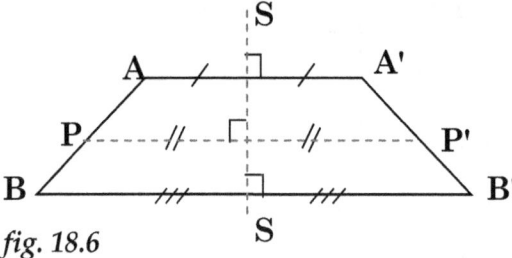

fig. 18.6

You should have noticed that if:-

(*a*) A figure has a line of symmetry S-S, therefore for every point, P, on the figure there is a corresponding point P', the image of P under symmetry about S-S.

(b) The line PP' is bisected at right angles by the line S-S. This means that the point P and P⊙are equidistant from the line S-S.

18.4 Symmetry in Coordinate System

We can have lines on the grid system working as lines of symmetry for some figures or points on the grid. Study the following figures carefully.

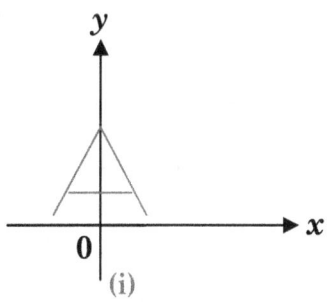

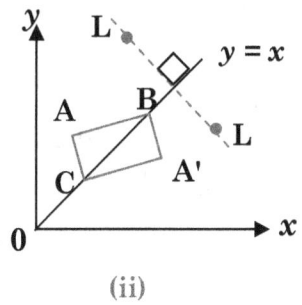

fig. 18.7

In part (i) of figure 18.7, we have the line $x = 0$ (or y-axis) dividing the letter "A" into two equal parts. In this case the line $x = 0$ acts as an axis of symmetry.

In part (ii) of figure 18.7, the line $y = x$, acts as the axis of symmetry. On the line LL⊘ point L⊘is the image of point L. Point A' is the image of point A. The points B and C are on the line of symmetry $y = x$ and are their own images. Therefore, points B and B'; C and C' are each pair at the same point.

Example 1

Given the line $y = x$ as the line of symmetry, what are the coordinates of the images of the points A, B, C and D?

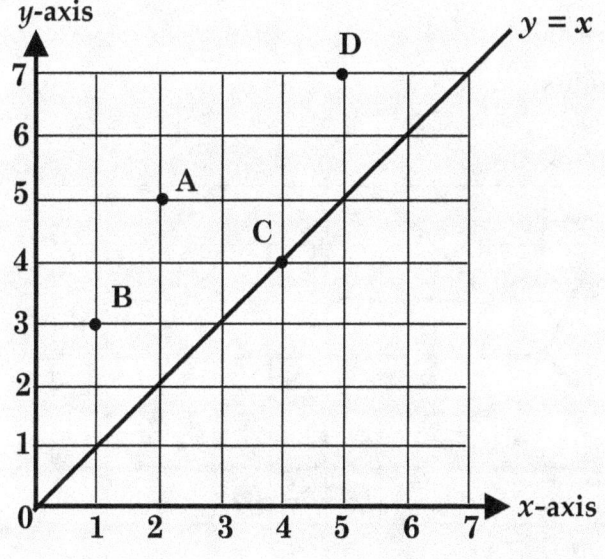

Solution

Remember that each point is the same distance from the line of symmetry as its image. When the point and its image are joined, the line joining them cuts the line of symmetry at 90°.

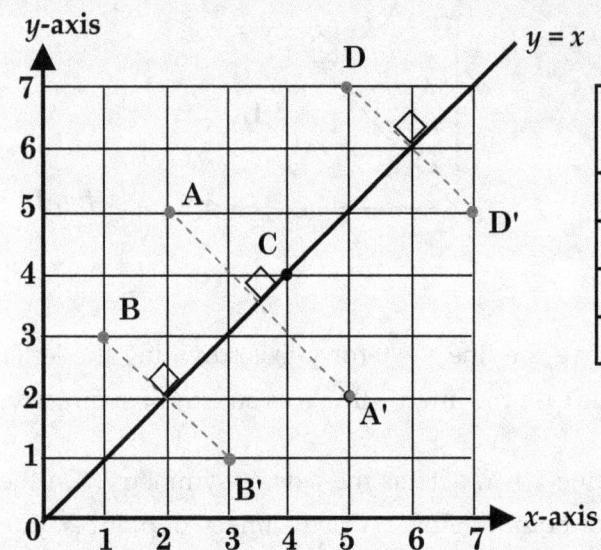

	Points coordinates		Images coordinates	
A	(2, 5)	A'	(5, 2)
B	(1, 3)	B'	(3, 1)
C	(4, 4)	C'	(4, 4)
D	(5, 7)	D'	(7, 5)

Example 2

Given the coordinates for figures as follows, draw the figure and their images with the line $y = 2$ as the line of symmetry.

  (i) A(-3, 4), B(-2, 5), C(-1, 4), D(-1, 3), E(-3, 3)

  (ii) F(0, 2), G(1, 4), H(2, 2), I(1, 1)

  (iii) J(3, 3), K(5, 5), L(6, 4), M(6, 3)

Solution

Images are shown in dotted lines

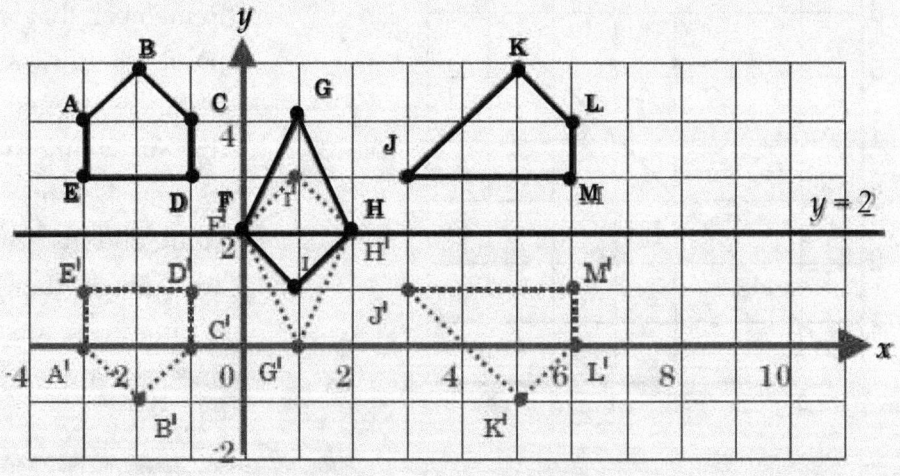

Exercise 18d

Given the lines of symmetry m-m on the grids below, find the coordinates of the images of the given points through the mirror lines.

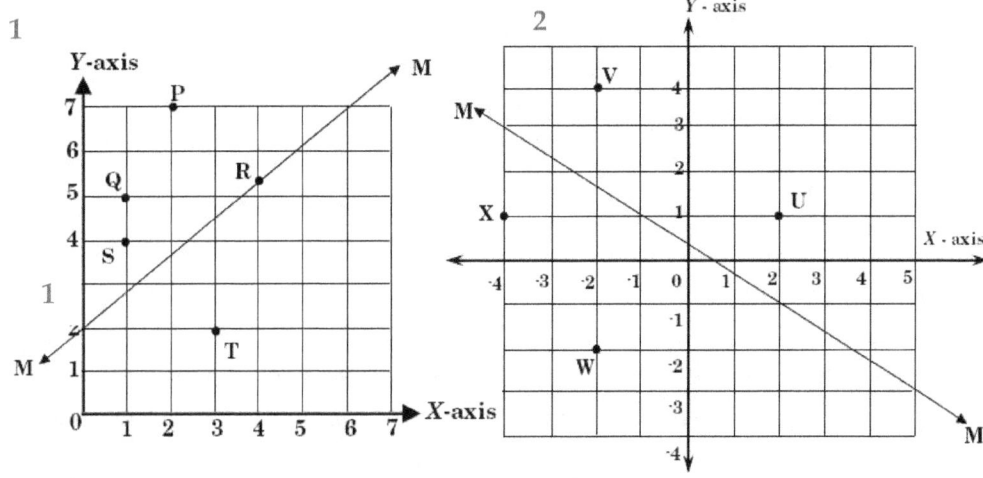

3

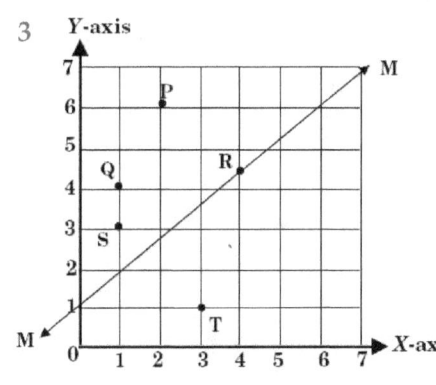

4 In each of the numbers 1 — 3, find the :

(a) coordinates of the points on the line of symmetry.

(b) equation of the line of symmetry.

Given the figures on the grids that follow, draw on the same grids the images of these figures through the line of symmetry m-m as given in each case.

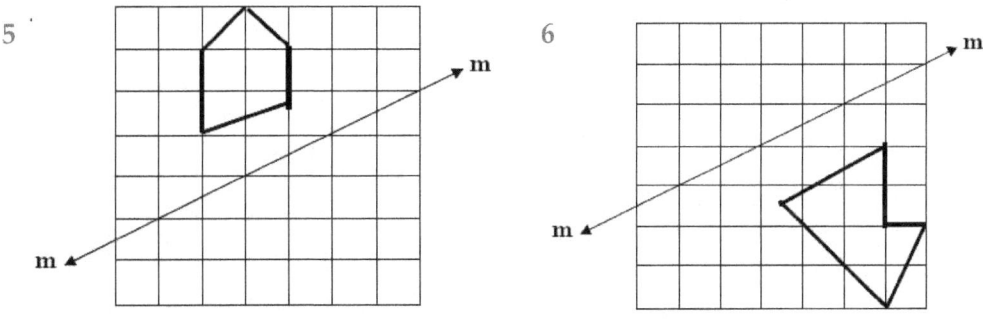

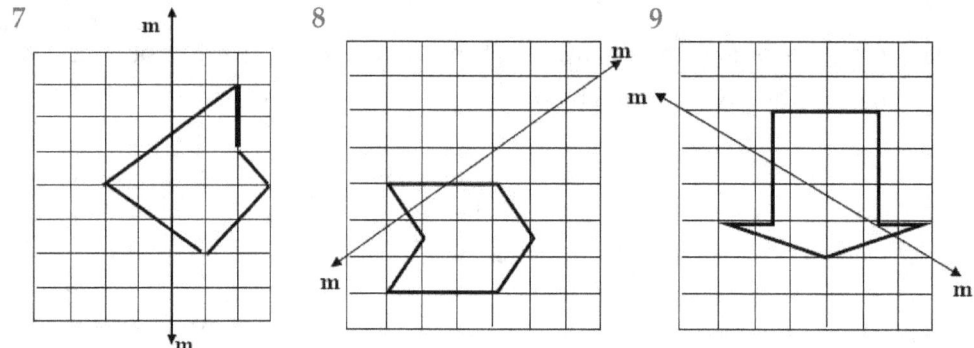

18.5 Reflection

When you view your image in a mirror, it will looks exactly like you, though there will be some differences. Touch your right hand side ear while viewing yourself in the mirror. What happens in the mirror? Try to move further away from the mirror. What happens to your image in the mirror? In reflection, the mirror acts as the line of symmetry. So, for reflection we shall have the object, the mirror (line of symmetry) and the image. Earlier in this chapter we looked at the figure being divided into similar parts, but in reflection we have the whole figure reproduced as an image. For reflection we have the following characteristics;

fig. 18.8

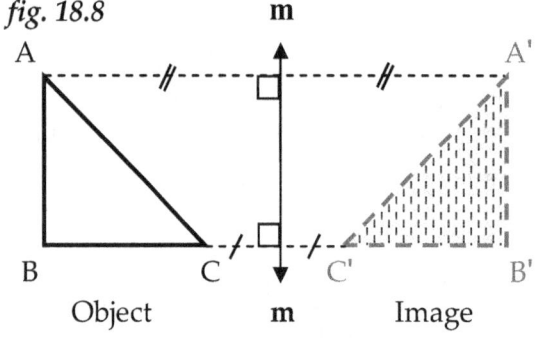

Object   m   Image

Activity 18.3

Given the object below and the mirror line mm, draw its image and label all the points on it.

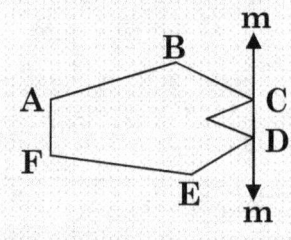

1 Every point on the object has a corresponding point on the image. For reflection under the mirror line m-m in fig.18.8 above; Point A corresponds to A', B corresponds to B', and C to C'.

2 The corresponding points on the object and image as in fig.18.8, point A and A', B and B' and C and C' are each pair equidistant from the mirror line. This means that the distance of point A from the mirror line is equal to that of its image, A', from the same mirror line.

3 The mirror line m-m is a perpendicular bisector of the line which connects the two corresponding points of the object and image. From fig. 18.8, line AA' and the mirror line m-m meet at 90°.

4 The shape and the size of the object and image are the same. We can therefore use a tracing paper, if we are given the object to obtain the image.

Example 1

Given the mirror lines and the object; draw their images.

(i)   (ii) 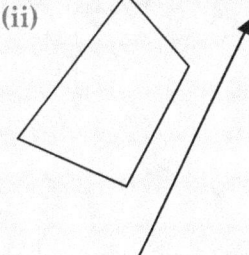   (iii)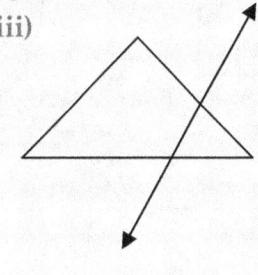

Solution

Remember that;

(i) the distance between the mirror line and the object is equal to the distance between the mirror line and the image.

(ii) all the corresponding points are equidistant from the mirror line.

(iii) the object and image have the same shape and size, but of inverse congruence.

(i) 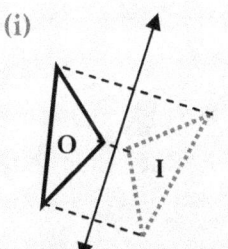   (ii) 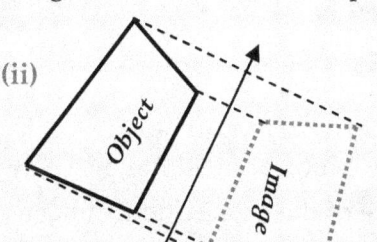   (iii)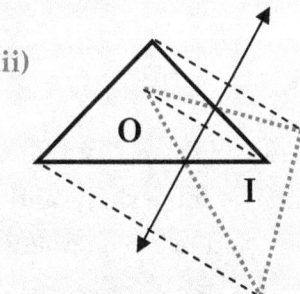

Exercise 18e

Through the mirror lines given, draw the images of the given objects (use tracing paper).

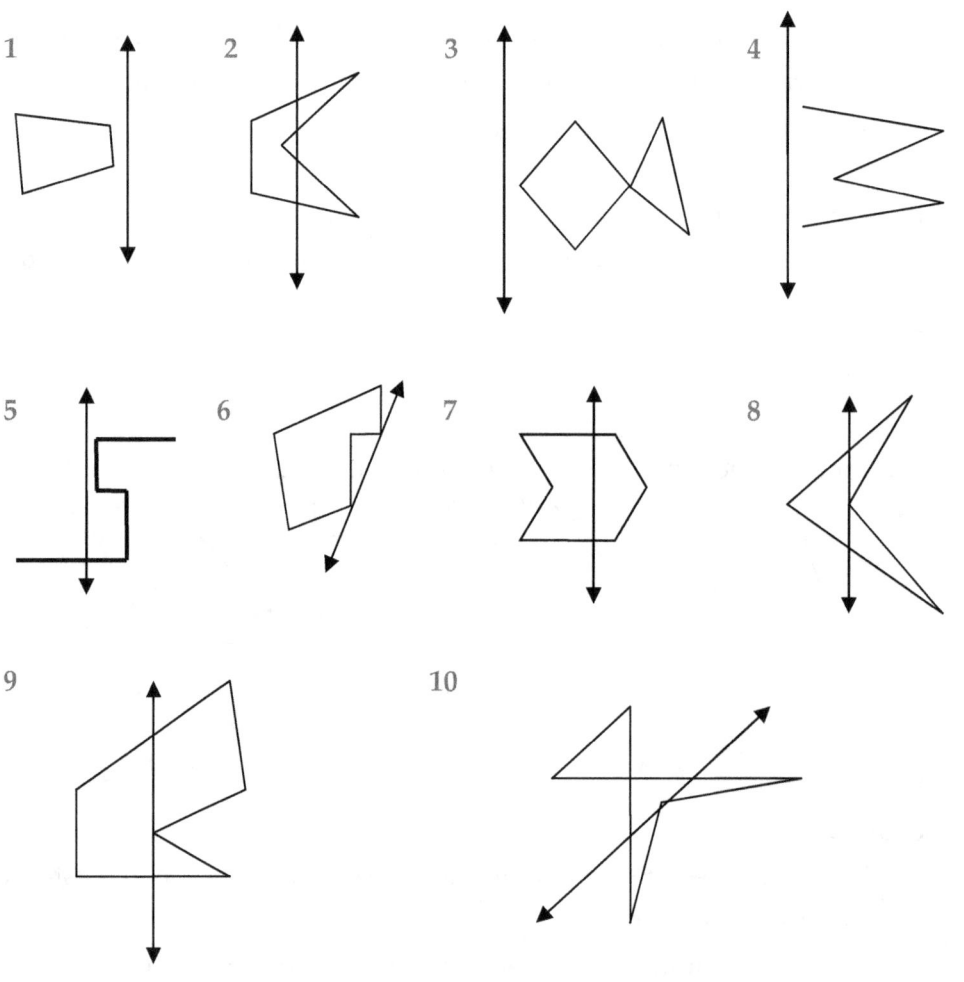

18.6 Congruence

For congruence of figures we look at the difference between the object and the image. If the shape and size of the object and its image do not differ then they are said to be congruent.

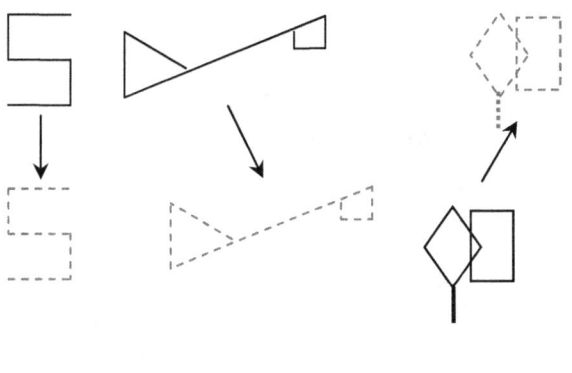

If two plane shapes are congruent and we can slide one so that it fits exactly on the other, the congruence is said to be *direct*, for example;

Slide the object to fit exactly onto its image in each case as shown in fig.18.9(a).

Fig.18.9(a)

And if the object fits onto its image only by turning it over, then slide and/or rotate, the congruence is said to be *opposite (or inverse)*, for example;

Turn over and slide the object to fit exactly onto its image in each case as shown in fig.18.9(b).

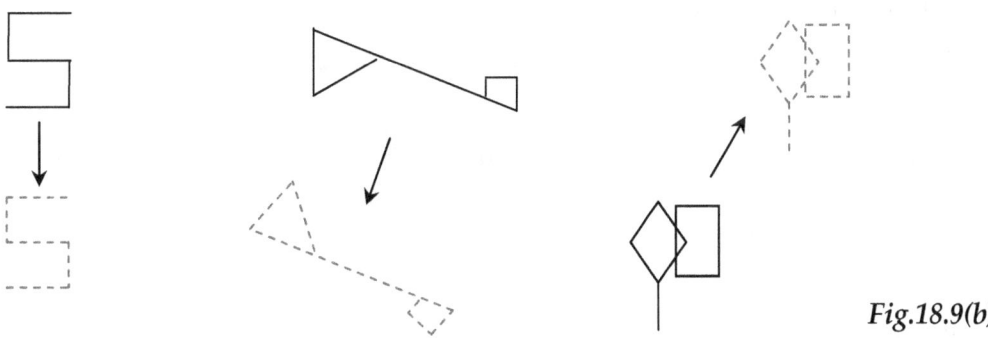

Fig.18.9(b)

Activity 18.4

1 Draw four pairs of figures showing direct congruence.
2 Draw four pairs of figures showing opposite (inverse) congruence.
3 On a card board or hard paper, mark and cut out the following shape, and use it to demonstrate direct and inverse congruence.

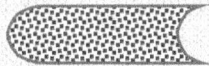

Example 1

In the figures below find the type of congruence in each case. The shaded figure is the image, whereas the unshaded is the object.

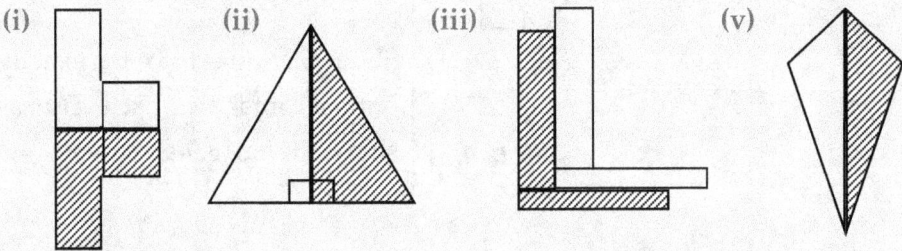

(i)   (ii)   (iii)   (v)

Solution

(i)   Inverse congruence, since it requires to first turn over then slide to fit object onto its image.

(ii) Inverse congruence, since it requires to first turn over then slide to fit object onto its image.

(iii) Direct congruence, since it requires only sliding to fit the object onto its image.

(iv) Inverse congruence, since it requires to first turn over then slide to fit object onto its image.

Example 2

In the figures below, find the type of congruence in each case. The images are shaded

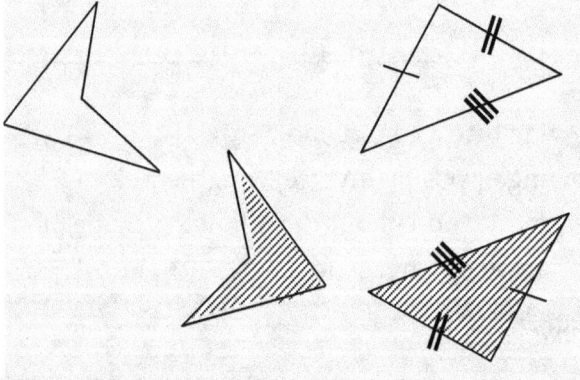

Solution

(i) Inverse Congruence, since we need a turn over then slide.

(ii) Direct congruence, since we only rotate a bit then slide.

Exercise 9e

1 Draw five pairs of figures which are directly congruent.
2 Draw five pairs of figures which are inversely congruent.

State whether the following figures are directly or inversely congruent.

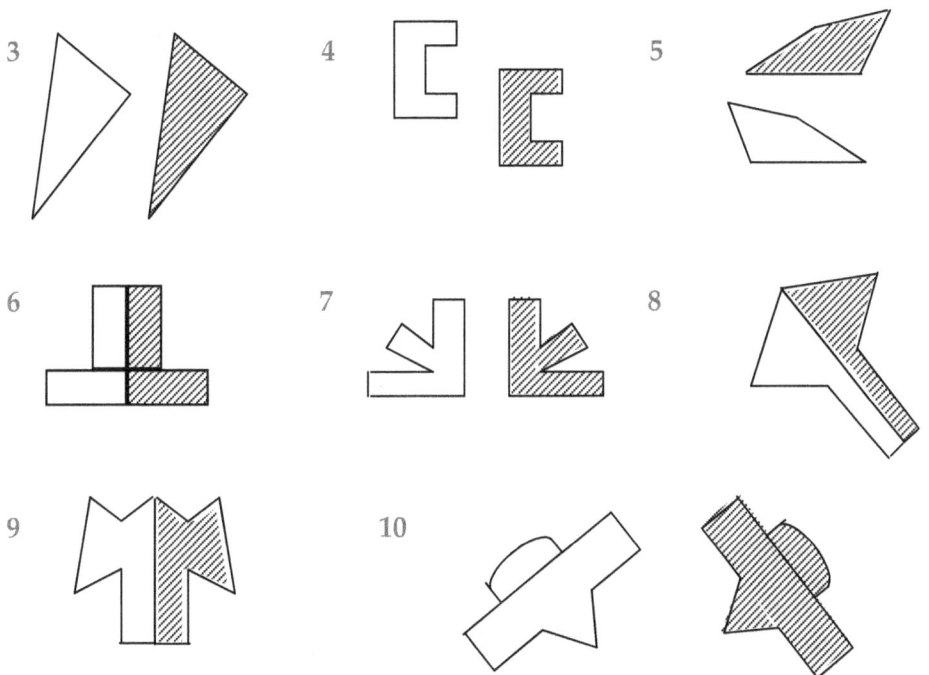

REVISION EXERCISE 18

Trace out the following figures, cut them out and try to fold them such that one flap fits exactly on the other. How many ways can you fold each of them?

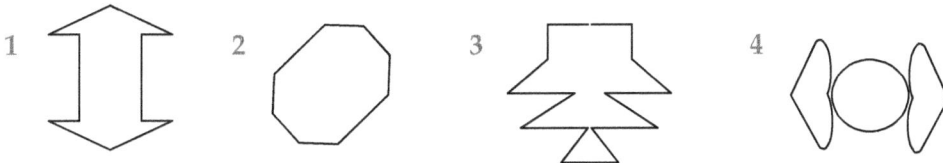

Use dotted lines to show the lines of symmetry in each figure:

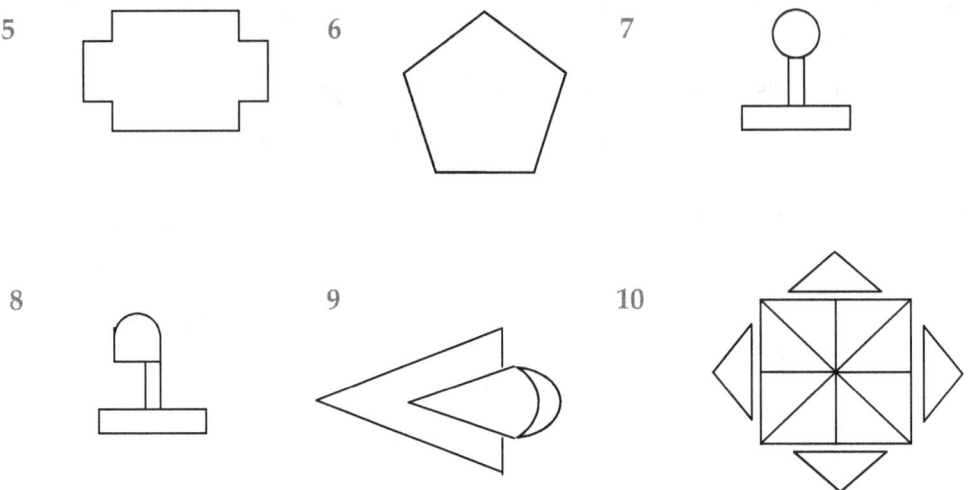

5   6   7   8   9   10

Given the grids below and the named lines of symmetry. Find and draw the images of these points. Give the coordinates of image points.

11

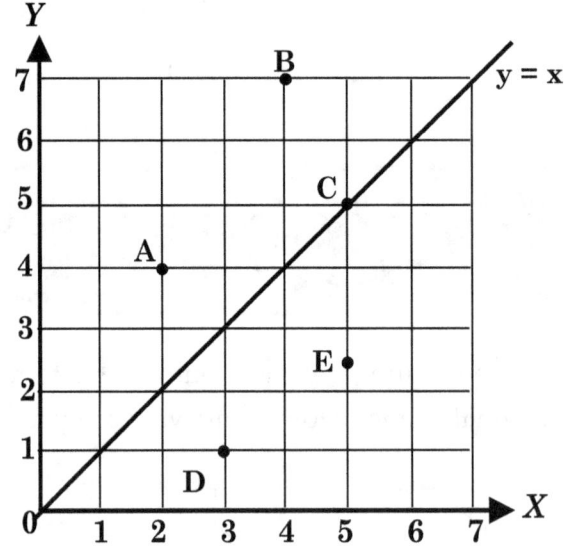

12

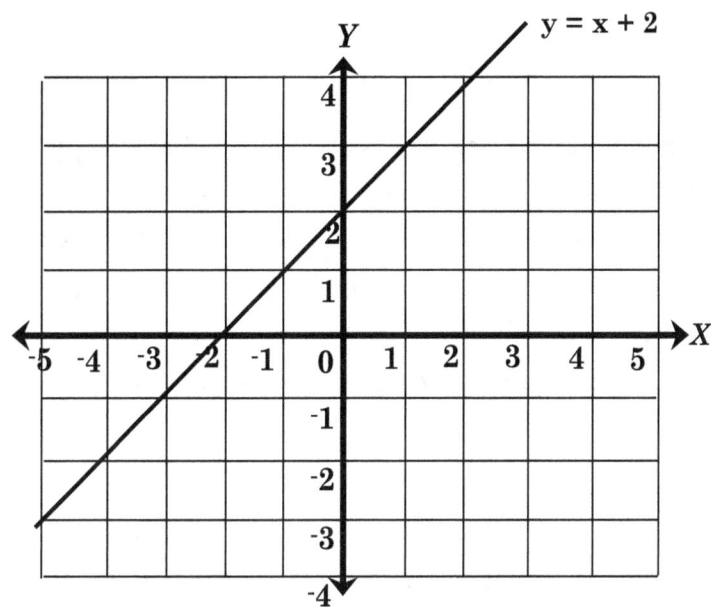

y = x + 2

Draw the images of the following figures through the mirror lines shown. (use dotted lines for images)

13

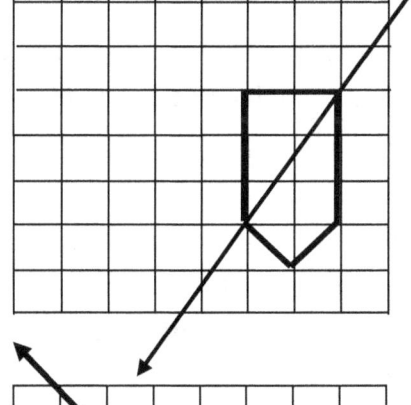

14

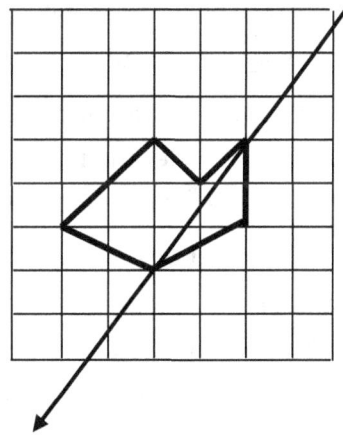

15

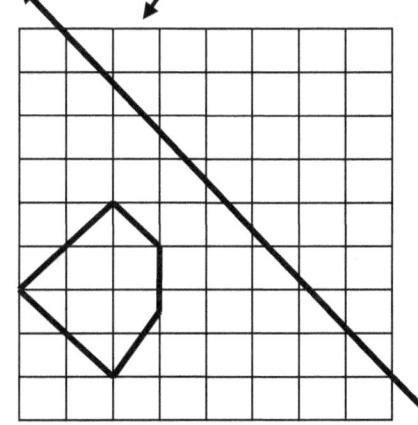

MODEL EXAMINATION TYPE TEST PAPERS

These model examination type test papers involve all what is required for real examinations in terms of scope and coverage, questions distribution, simplicity/complexity, timing and objectivity; therefore should be handled as and under similar conditions as real examinations. The following instructions apply to each of the following test papers.

TAKE ABOUT 4 — 5 MINUTES READING INSTRUCTIONS.

Instructions

- The paper is composed of two sections A and B. Section A contains ten (10) short questions equivalent to 40 marks. Section B contains five (5) questions with each equivalent to 15 marks. Each paper takes 2 hours.

- Attempt ALL Questions in section A and any FOUR(4) from section B. This leads to the total mark of the paper to be out of 100marks (40 marks for section A and 60 marks for section B).

- Where an extra question is given you can attempt it, if you feel you can do it much better than any of the four numbers you have attempted in section B. In this case you do five questions and the best four are considerd.

- In calculations, you are advised to show all the steps in your working. No mark is awarded for a correct answer with wrong method/working, however one can earn a mark (s) for the correct working even if the answer is wrong. So be more careful with your working rather than only thinking about the answer you   intend to get.

- You may start with any number in the paper as long as it is correctly labeled, and avoid mixing numbers of different sections or scattering numbers of the same section.

- Make sure your work is neat. Do rough work on a separate piece of paper. If you make a mistake avoid multiple crossings; just cross out with a single line running from the top right to the bottom left and mark *"Omitted"*.

MODEL TEST PAPER 1

S E C T I O N   A   (4 0   M a r k s)

1. Simplify   (i) $\dfrac{(^-2) + (^-6) \times ^-7}{^-40 \text{ of } 2}$ (ii) $12x^2yp^3 \div 8p^4xy$

2. Find the LCM and HCF of 48, 120, 90 and 24.

3. List down all the odd numbers which are multiples of 3 and are square numbers between 1 and 100.

4. Given the number 48224, is it exactly divisible by:
   (i) 3   (ii)   4   (iii)   5   (iv)   11   (v)   12

5. Fill in the missing numbers:
   (i) 1, 3, 6, 19, ….., ….., ……   (ii) -10, -7, -4, ……, ….., …..

6. Work out: $\dfrac{3^{11} \times 3^{-4} \times 9}{27}$ using indices

7. List down all the composite numbers between 0 and 20, that are sums of 2 consecutive prime numbers.

8. Change the following: (i) 5% to a fraction   (ii) 0.61 to a percentage

9. Expand: $(3 - x)(2y + 1)(x + 2)$

10. Convert the decimal 0.232323…, to a fraction

S E C T I O N   B   (6 0   M a r k s)

11. (a) Given the figure below, find the value of x, y and z.

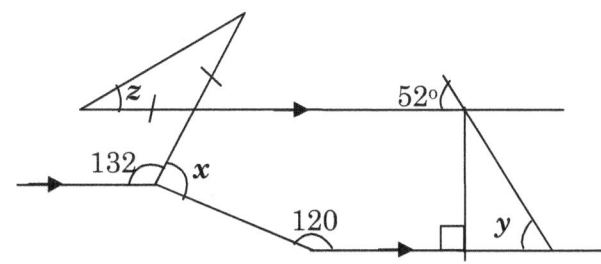

(b) Find the perimeter of the figure below.

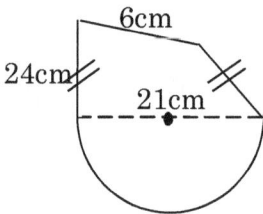

12. Given the equation of the line $2y = x - 5$.

    (a) Tabulate the coordinates of this line such that $^-5 \le x \le 4$.

    (b) Draw this line on a grid. Choose a suitable scale.

13. (a) The length of the road is 6cm on the map. If the scale used to draw this map is 1 : 500,000, what is the actual length of the road?

    (b) State whether the triangles ABC and LMN are congruent. If they are, state the type of congruence they have.

14. (a) Solve for x in

$$\frac{3x - 2}{4} = \frac{x - 1}{3}$$

    (b) Work out: $1\frac{1}{4} \div \frac{1}{8} \times 3\frac{2}{5} + \frac{1}{3} - 1$

15. Given the Venn diagram, answer the following questions.

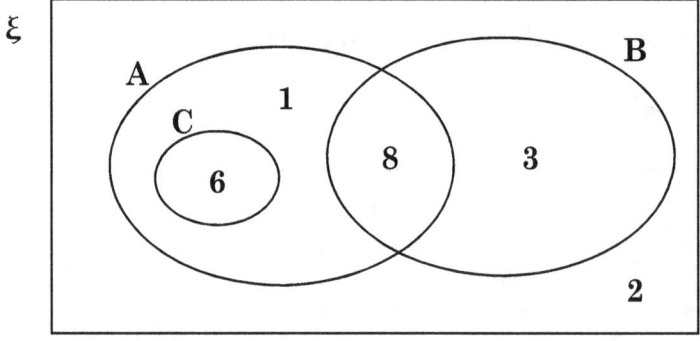

(a) Find ;

(i) $n(\xi)$

(ii) $n(A)$

(iii) $n(B)$

(iv) $n(B \cap C)$

(v) $n(A \cup B \cup C)^{l}$

(vii) $n(A \cap C)$

MODEL TEST PAPER 2

SECTION A (40 Marks)

1. Evaluate: (i) $\dfrac{0.06 \times 7.61 + 34.01}{34.4666}$ (ii) $164 - 32.416$

2. (*a*) Express 144 as prime factors

   (b) Write 36.4006 correct to the nearest thousandths.

3. (*a*) Given set A = {1, 2, 3, 4, 5, 6} and Set B = {2, 3, 5, 7}.
   Find   (i) n(A ∪ B)   (ii) n(A ∩ B)

   (b) List all the subsets for set P = {1, 2, 3}

4. Tabulate the coordinates on the line $y = 2x - 1$ for $^-8 \le x \le 0$.

5. (*a*) How many lines of symmetry has the figure to the left got?
   (b) If a regular polygon has an exterior angle of 45. Find its interior angle sum.

6. How many cubic meters are there in 0.0001km^3.

7. In a class of 40 students, an 8th are absent, $2/5$ of the class are present but outside the class and the rest are in class. How many students are inside the class?

8. Simplify: (i) $\dfrac{3(2x + 3)(x - 2)}{3x - 6}$ (ii) $\dfrac{2x}{5} - \dfrac{3x}{4} \times 2\,^6/_7$

9. (a) Express 2.41 hours to minutes

   (b) If set P = {numbers which are neither prime nor composite}, List the elements of this set

10.

   Find the value in degrees for w, x and y; giving reasons why?

SECTION B (60 Marks)

11. (a) (i) List the first 10 even 10 odd numbers.

  (ii) Add the first even numbers to the first odd, number the second to the second, the third to the third, and soon.

  (iii) Comment about the obtained series.

 (b) Express $^{11}/_{12}$ as a decimal

12. Given the grid in the figure that follows answer the questions that follow;

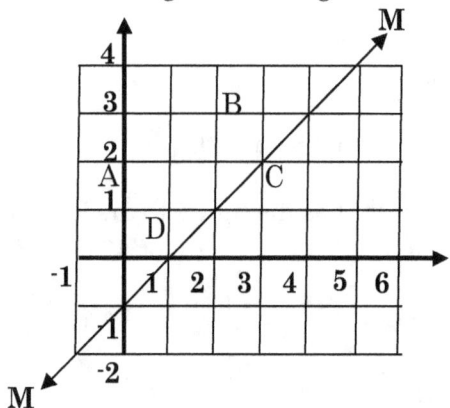

If the figure ABCDE undergoes reflection through the mirror line m-m;

  (i) Find the coordinates of the images of points A, B, C, D and E and show them on the same grid.

  (ii)   Join the points A¹, B¹, C¹, D¹ and E¹ to obtain the image of ABCDE.

  (iii)   Find the equation of line M-M.

  (iv)   Comment about point C.

13. In a class of 49 students, 25 like volley ball (V), 12, like basket ball (B) and the member of students who like neither of the two games is less than twice those who like both games by 1.

 (a) Represent this information on a Venn diagram

 (b) Find   (i) the number of those who like both games

              (ii) the number of those who like neither of the games.

14.   (a) Work out: $1\,^2/_7 \times {}^{14}/_{17} - {}^4/_5 \div 2$

        (b) Complete: $^2/_5, {}^8/_{20}, {}^{32}/_{80}, \ldots\ldots, \ldots\ldots, \ldots\ldots$

15. In making music, a bass drum is hit after every 26 seconds, a clap is made after every 54 seconds and a whistle is made after every 72 seconds. If they are all started at the same time, after how many seconds will they be struck again at the same time?

Extra Question

16. Copy and complete the following table

Water consumed in one month by 3 different homes						
Previous reading	Current reading	Units used	Cost per unit	Service fee per month	VAT rate	Amount due
11801	11836	_____	Sh. 3,2500	Sh. 3,000	18%	_____
_____	01028	102	Sh. 2,000	Sh. 2,200	16%	_____
08460	08614	_____	Sh. 4,000	Sh. 2,000	18%	_____

MODEL TEST PAPER 3

SECTION A (40 Marks)

1. (a) Express 0.062m to mm.

   (b) Is the number 292392 divisible by 12? Explain your answer.

2. (a) Add: 240 + 0.418 + 1.001.

   (b) A student obtained 46 marks out of 115marks. Express his mark as a percentage.

3. Find the area of the following figure.

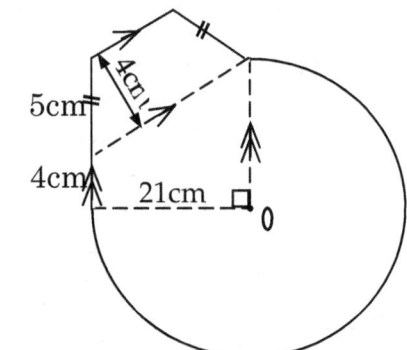

4. (i) Expand: $(2x - 3p + 1)(31 - xty)$

   (ii) Factorise: $3x^2pq^3 + pm^2q^3 + 6x^2y + 2m^2y$

5. (a) Find the values of a and b in degrees.

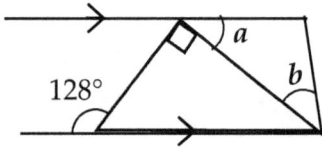

   (b) A number is multiplied with its reciprocal and the result is squared, what is the final answer.

6. The distance on a plan is 20 mm if the actual distance on ground is 2m. What scale was used?

7. A sector has a radius 315 cm and an angle of $32°$ at the centre. Find its arc length.

8. Solve for x in: $2\frac{1}{2} - \frac{3x}{2} + \frac{4}{7} = \frac{10x}{7}$

9. Find $\frac{2}{5} \times \frac{3}{8}$ by illustration on diagrams

10. Express 2.31555… as a fraction.

SECTION B (60 Marks)

11. Given the sets   A {1, 2, 3, 4, 5, ... }   B = {0, 1}

                 C = {4, 6, 8, 9, 10, 12}   D = {2, 3, 5, 7}

                 E = {3, 6, 9, 12, 15}

(a) Name all the sets A, B, C, D and E

(b) State true or false   (i) $E \subset A$ (ii) $n(C \cap E) = 3$

                       (iii) $A \cap B = \{ \}$ (iv) $n(E) - n(D) = 1$

                       (v) $n(c) = n(E)$

12. (a) A pentagon has angles $21°$, $x°$, $3x°$, $(2x + 3)°$ and $(x - 2)°$

     (i) Find the value of x (ii) Give the sizes of all the angles in degrees

   (b) Arrange the digits 2,3,4,6 and 7 to obtain.

     (i) The greatest number divisible by 11

     (ii) The least number divisible by 11

     (iii) the greatest number divisible by 4.

13. (a) Given the tabulated coordinates below;

x	0	1	2	3	4
y	-4	-3	-2	-1	0

(i) Plot these coordinates on a grid.

(ii) Find the equation of the line.

   (b) How can you find triangular numbers? List the first ten triangular numbers.

14. (a) Given figures below,

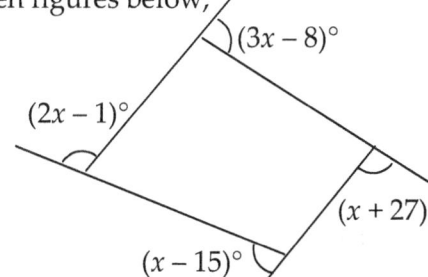

(i) Find the value of y

(ii) Find all the exterior angles.

$(3x - 8)°$

$(2x - 1)°$

$(x + 27)°$

$(x - 15)°$

(b)   Given the points A (5, -3) and B(8, 2), find the length of line AB.

15. Using a tracing paper, find the image of the figure below through the mirror line M-M.

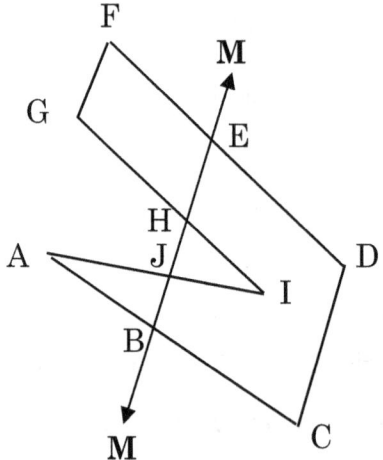

Extra Question

16. Two cross roads intersect at 90°. Mujuzi and Katamba started their   journeys from the road junction and each of them took a different road. After an hour, Mujuzi had covered 8 km and Katamba had covered   15 km.

   (a) If the road Mujuzi took was continuing in the eastern direction;

      (i) What are the possible road directions Katamba took?

      (ii) What was the bearing of Katamba from Mujuzi after an hour of traveling?

   (b) What was the shortest distance between Mujuzi and Katamba after an hour of travelling?

MODEL TEST PAPER 4

SECTION A (40 Marks)

1. Arrange the following in ascending order

   (a) $1/3, 2/7, 5/8, 7/9, 1/2, 5/7$ (b) 0.23, 2.3, 0.0023, 23.0, 0.00023.

2. Round the following to the nearest hundredths (i) 0.008294   (ii) 24.48167.

3. Work out: (i) $\dfrac{6 \times (125)^3}{5^2 \times 36}$ (ii) $\dfrac{-2 \times -9 \times -3}{36 \times -2}$

4. Use the number line to evaluate   (i) $-2 + (-3) + 4$ (ii) $-9 - 1 + 11 - (-2)$

5. I save $2/5$ of my monthly salary, spend $1/8$ of the rest on food and I spend the remaining on clothing. If I earn, shs. 180,000 per month, how much money do I; (i) Save   (ii) Spend on clothing   (iii) Spend on food

6. (a)   Prime factorise 2484.

   (b) The sum of 4 consecutive even numbers is 36. what are these even numbers?

7. Given that set   P = {all prime numbers less than 100}

                          Q = {all odd numbers less than 100}

   (a) List the members of each set.

   (b) Find (i) n(P)   (ii) n(Q)   (iii) n(P ∩ Q)

   (c) Is P ⊂ Q ? (Explain your answer)

8. Given points P $(4 1/5, -2 1/2)$ and Q $(1, -2 1/2)$, find the length of the line PQ.

9. Given the figure below, find its area.

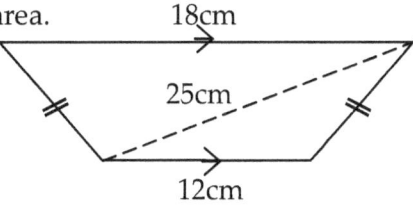

18cm

25cm

12cm

10. Evaluate: $\dfrac{1^3/_7 \div 5/_7 - 4/_7 + 3/_8 \times 4/_9}{79/_{70} + (3/_5 \text{ of } 7/_9)}$

SECTION B (60 Marks)

11. (a) Given the figures below; find the values of the angles represented by letters a, b, c, d and e and give reasons.

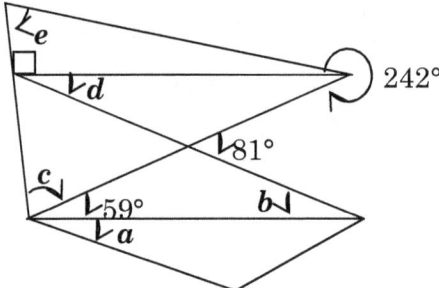

(b) The area of a piece of land is 0.0012 km². Express this area in; (i) Square meters

(ii) Square hectometres

12. In a class of 52 students, 30 like maths (m) and 32 like English (e) and 2 like neither of the two subjects.

    (a) Represent this information on a venn diagram.

    (b) How many students like both subjects?

    (c) How many students like only one subject?

13. (a) Given that $a = 2$, $b = {}^-3$ and $c = 6$, evaluate;

    (i) $\dfrac{3b - c}{6a}$
    (ii) $\left[{}^-2b + \dfrac{bc}{a}\right] abc$

    (b) Simplify: $\dfrac{mn\,(3b - pq)\,(4 - p)}{(p - 4)\,(pq - 3b)^2}$

14. On a grid such that x-coordinates are ${}^-6 \le y \le 6$ and y-coordinates are ${}^-6 \le y \le 6$, draw the following lines. (i) $2x - y = 0$ (ii) $4 - 3y = 2x$.

15. In a certain club there are $(y - 3m)$ men and $(3y - 11)$ women.

    (a) (i) Write an expression for the total number of members in the club.

    (ii) If there are more women than men in the club, write an expression for the number of women who exceed the men.

    (b) If the number of women in the club is 322 and exceeds the number of men by 223, find: (i) the values of y and m

    (ii) the total number of members in the club.

MODEL TEST PAPER 5

S E C T I O N   A   (4 0   M a r k s)

1. (a) List the multiples of 4 that are square numbers between 0 and 200.

   (b) Simplify:
   $$\frac{2^3 \times 113^\circ \times 8}{2^{-1} \times 2^5}$$

2. (a) Reduce 240 by 12%

   (b) Simplify: $\dfrac{3bx - 3blm - pqx + pqlm}{xy(3b - pq)(x - lm)}$

3. (a) Write a word phrase for: $n^2 - (^1/_n)^2$.

   (b) Given the digits, 1,2,4 and 8 form all the possible numbers divisible by 4 from these digits each used only once in a given number.

4. (a) Two angles are both congruent and complementary. What is the size of each angle?

   (b) Give the values of angles represented by the letters in the figure.

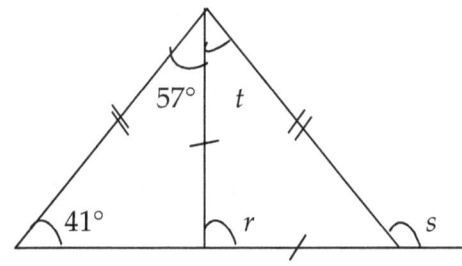

5. The actual length of a tree is 12m, by using a scale of 1:100, how many millimetres will I use to represent this height on a paper?

6. The circumference of a circle is 396 cm, what is its diameter.

7. Work out: (i) $\dfrac{42.48 \div 0.04}{40} \times \dfrac{4.2}{106.2}$ (ii) $1\frac{1}{2} - ^3/_7 + 3$

8. Given the figure,

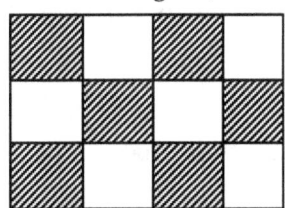

   (i) Find the fraction for the shaded part.

   (ii) What percentage of the figure is shaded?

9. A polygon with five sides has its exterior angles in which each exceeds the following by 2°. Find the exterior angles of this polygon.

10.

If AC = 27.43cm, find lengths:

(i) BE

(ii) AB

(iii) AE

(iv) AD

SECTION B (60 Marks)

11. (a) In a class of 48 pupils 4 grow both bananas and coffee, 8 grow neither of the two crops. The number of those who grow bananas only is 3 times the number of those who grow coffee only. Find;

(i) The number of those who grow coffee only.

(ii) The number of those who grow bananas only

(iii) What fraction is the number of those who grow both.

(b) Represent the region $(A \cup B \cup C)^{I}$ on a venn diagram by shading.

12. (a) My age is $^2/_3$ of my sister's age, my sister's age is 1/5 of our father's age doubled. If our father is 60 years find:

(i) how old is my sister?

(ii) how old I am ?

(iii) an eighth of the sum of our ages (i.e., father, son, and daughter).

(b) (i) Out of 150 marks a boy obtained 98, what was his percentage marks?

(ii) A number is reduced by its reciprocal and the result is 0, what is the number?

13. Given the venn diagram next page, answer the questions that follow.

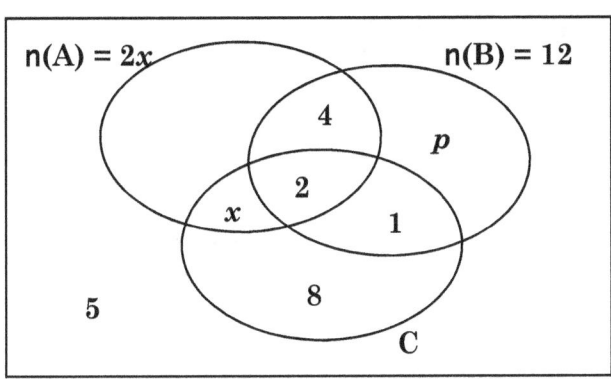

Find: (i) the value of p

(ii) the value of x

(iii) n(A)

(iv) n(C)

(v) n(E)

(vi) n(A ∩ B ∩ C)^{l}

(vii) n(A ∪ B ∪ C)^{l}

(viii) n(A ∩ C).

14. From the figure below find the sizes of angles represented by letters in degrees.

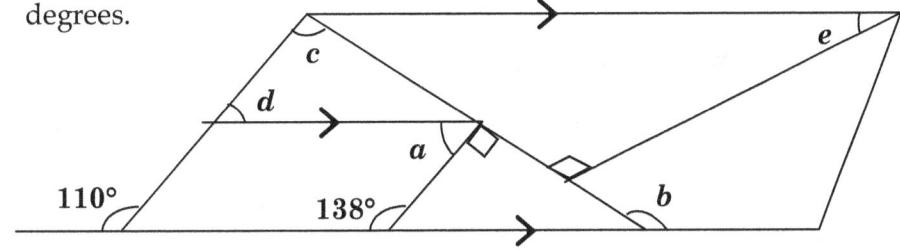

15. If I bought $(3y + 25)$ books at shs. $(x - 500)$ each.

   (a) Write expressions for

   (i) The total amount I paid for these books

   (ii) My balance after paying, if I had shs.$(4200 - x)y$ altogether originally.

   (b) If x = 1000 and y = 15 ;

   (i) What is the price of one books?

   (ii) How many books did I buy?

   (iii) How much did I pay for all the books?

   (iv) What was my balance after paying for the books?

MODEL TEST PAPER 6

1. Factorise;   (i) $2xy + 4xp - p - 2p$ (ii) $y^2x - x^2y$

2. Simplify: (i) $\dfrac{2^{-1} \times 3^2 \times 2^4}{3^{-3} \times 2^2}$ (ii) $\dfrac{mn}{2p} - \dfrac{3mn}{i} \div \dfrac{mn}{2pl}$

3. (a) How many times does $^1/_8$ go into $^3/_4$?
   (b) What percentage of 25 is 6?

4. Fill in the missing numbers.   (a) -9, -3, ..., ..., 15, ...

     (b) ......, ..., ..., 2, -4, 8, -16.

5. Express the following as decimals: (a) $^1/_8$ (b) $^2/_{11}$

6. If set V = {a, e, i, o, u} and set C = {a, b, c, d, e}.   Find: (i) V ∩ C   (ii) V U C

7. (a) List all the factors of 330.   (b) What is the H.C.F of 11, 55 and 121.

8. (a) Find the distance of line AB, where A(-3, 1) and B(2.5, -5)

   (b) Express 9 out of 15 as a percentage.

9. A man packed tins of sugar of weight 520 grams, 420 grams, 500 grams and 360 grams.
   (a) Find the average weight for these four tins
   (b) If he packed this sugar into 5 tins of equal weight, how heavy will each be?

10. The area of a piece of land on a map is 24 cm². What is the actual area of this piece of land if the scale for the map is   (i)   1 : 10,000

                                                                        (ii)   1 : 500,000

11. Given sets   L = {all multiples of 5 less than 90}

                 M = {all odd multiples of 5 less than 90}

N = {all multiples of 10 which are divisible by 4 and less than 100}

P = {all multiples of 3 which are divisible by 5 and less than 100}

    (a) List down all the members of sets L, M, N and P

    (b) State *true* or *false*.   (i) M ⊂ N   (ii) N ⊂ P

                           (iii) n(N) = 4   (iv) 100 ∈ N

    (c) Find L ∩ M ∩ N ∩ P.

12. (a) On a grid draw the following lines

        (i) AB , where A (0, 2) B (-2, 4)

        (ii) CD , where C (-4, 2) and D (1½ , -5)

    (b) Find the lengths of AB and CD.

    (c) Plot X(-1.5, 2) and Y(1.6, 3) on the same grid.

    (d) Jon X to Y and find length XY

13. A man earns shs. 8400 per month. He used ¹/₈ of his salary on food, ²/₅ of his salary goes on rent, he spends ½ of what he uses on rent on clothing and he saves the rest.

    (a) Find how much he spends on: (i) Food   (ii) rent

                                       (iii) clothing   (iv) saving.

    (b) What fraction of his salary does he spend on clothing?

    (c) Express the money he spends on rent as a percentage of his monthly salary.

14. Kikomeko had 92 full pieces of chalk in a box and as he put them in the shelves 43 pieces got broken and were thrown away. He then added 87 full pieces to the box. He then supplied the pieces from the box such that S.1 obtained 16 pieces, S.2 obtained 22 pieces, S.3 obtained 31 pieces and S.4 obtained 44 pieces.

    (a) How many full pieces did Kikomeko have in the box before the supply?

(b) How many full pieces remained in the box after the supply?

(c) If he was to give out the remainder to all the classes equally, what would be the final number of pieces per class?

(d) If all the four classes were to share all the pieces of chalk equally, how many pieces would each class get?

15. In the figures below, use a tracing paper to draw the images of the given figures

(a) M

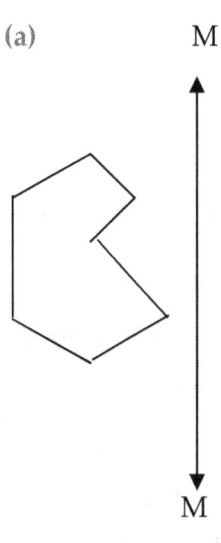

M

(b) M

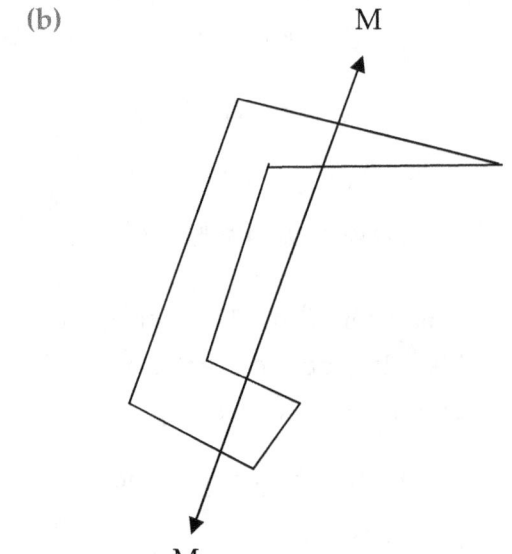

M

Extra Question

16. (a) A lorry can make a journey at 80km/hr in 3 hours. How long will the same journey at;

    (i) 50km/hr   (ii) 72km/hr   (iii) 120km/hr?

(b) 4 men can build a room in 2 days, find;

    (i) how many men will build the room in ½ day.

    (ii) the time 1 man takes to finish one room.

MODEL TEST PAPER 7

S E C T I O N   A   (4 0   M a r k s)

1. Work out   (i) $4 - 3.614$ (ii) $0.00081 \times 20,000$

2. (a) How many sub sets has the set P = {a, b, c, d} got ?

   (b) List all the subsets of set P.

3. Solve for x in: $3x/7 - 5 = 2\frac{1}{4} - x/2$

4. Use indices to work out: $\dfrac{216 \ \times \ 81 \ \times \ 42}{7 \ \times \ 16 \ \times \ 243}$

5. Express: (a) 42.1 minutes as seconds

              (b) 3.14 mm^2 to centimetres squared.

6.
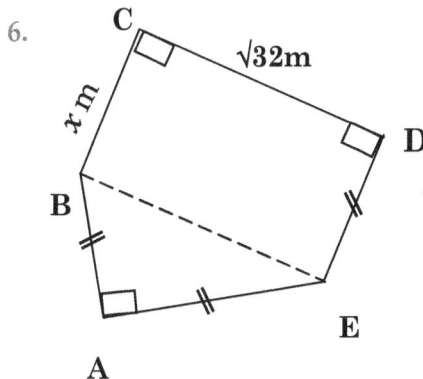

    (a) Find the value of x in the figure on left.

    (b) Find the area of ABCDE.

7. Work out: $\dfrac{2\frac{1}{2} + 4\,^3/_5 \ - \ ^7/_8}{^1/_3 \ \text{of} \ ^{40}/_{83}}$

8. Draw the line $2y = 6x - 1$ on a grid for $^-4 \le x \le 5$.

9.
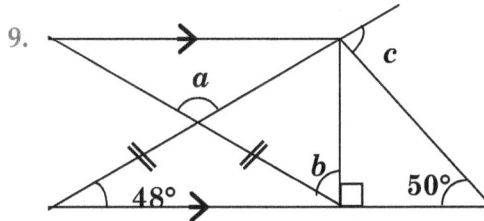

Find the values of a, b and c in degrees.

10. (a) Two angles of a right angled triangle arc complementary, if one is 4 times the other, what are the sizes of these angles?

    (b) The three internal angles of a scalene triangle are $(x - 3)°$, $(2x + 7)°$ and $5x°$.

    Find: (i) the value of x (ii) the size of each of three angles in degrees.

S E C T I O N   B   (6 0   M a r k s)

11. Given sets: $\xi = \{1, 2, 3, 4, 5, 6, 7, 8, 9, 10\}$ $E = \{2, 4, 6, 8, 10\}$
    $P = \{2, 3, 5, 7\}$ $T = \{1, 3, 6, 10\}$

    (a) Represent these sets on a venn diagram

    (b) Find: (i) $E \cap T$ (ii) $n(E)$ (iii) $n(E \cap P \cap T)$ (iv) $n(E \cap P \cap T)^I$

    (c) List the numbers which belong to at least two of the sets E, P or T.

12. (a) On a grid plot the following points A($^-$3, 2), B(0, 5), C(2, 3) and D($^-$1, 0)
    (b) Join the points A to B, B to C, C to D and D to A
    (c) What is: (i) the name of the figure ABCD drawn?
                        (ii) the perimeter of figure ABCD?
                        (iii) the area of the figure ABCD?

13.

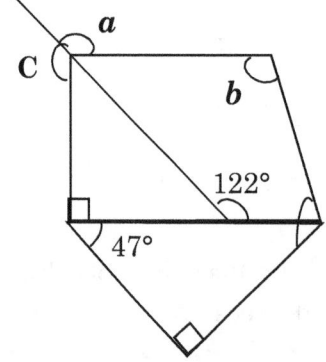

    (a) Find the value of angles marked by letters in degrees.

    (b) Express 0.213 as a fraction

14.   (a) Use a number line to find: (i) ⁻2 + (⁻3) + 5 (ii) 7 + (⁻8) + 1

(b) What is the lowest number that is divisible by 20, 42 and 6 ?

(c) Given the digits 4, 15 and 8.

(i) What is the difference between the largest and the smallest number divisible by 11 which you can obtain by arranging these digits without repeating any?

(ii) Is the difference divisible by 11? If it is, divide it and give your answer.

15. (a) State the types of congruence for the following figures.

(i)   (ii)

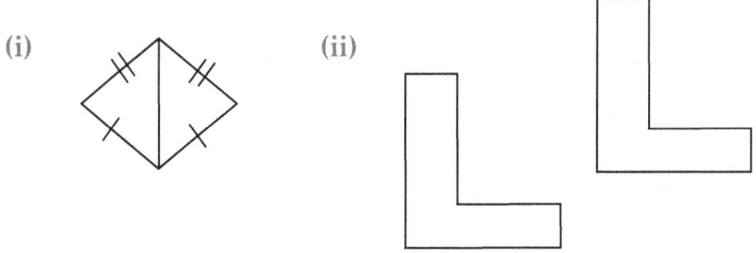

(b) Work out:

$$\frac{^{-}1\!/_2 \; + \; ^2\!/_5 \; + \; ^2\!/_3 \; \div \; ^{17}\!/_{45}}{(^1\!/_3 \; - \; ^1\!/_8) \; \text{of} \; 2\,^2\!/_5}$$

(c) Write an expression for the phrase "A number cubed is reduced by five times the number."

(d) A set has 9 members, how many sub sets can you form from it?

MODEL TEST PAPER 8

SECTION A (40 Marks)

1. Re arrange the following in ascending order:

   ½, 0.6, ³/₅, 0.4, 0.75, ⁵/₈, 35%, 65%

2. (a) Find the square root of the reciprocal of 25.

   (b) Reduce 3 by its reciprocal.

3. The area of a square is 20.25 cm² find its;   (i) perimeter

   (ii) diagonal length

4. The diameter of a bicycle wheel is 84cm. if the bicycle travels a distance of 1.6km, how many revolutions will the wheel have made?

5. Work out: $(^3/_4 \div {}^5/_{12} - {}^1/_3)$ of $^{15}/_{22}$

6. Approximate the number 344.8261 to:

   (i) The nearest whole number   (ii) The nearest hundredths

7. (a) How many lines of symmetry has a circle?

   (b) Express ⁵/₈ as a percentage.

8. 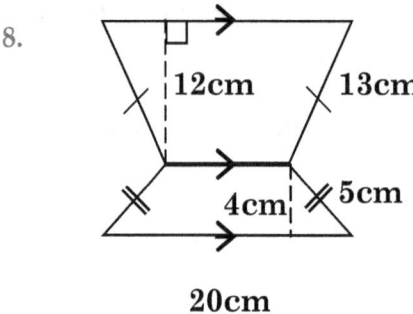 Find the perimeter and area of the figure below;

9. Solve for x in: $\dfrac{2x - 5}{7} = {}^1/_3 + {}^2/_5$

10.

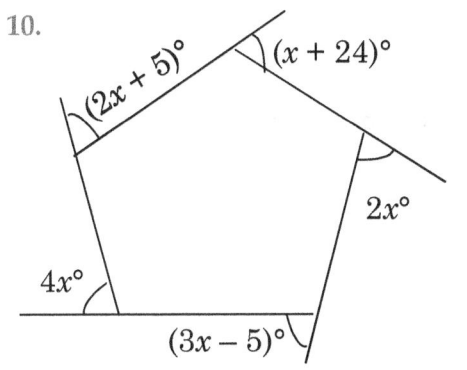

In the figure on the left, find:

`(i) the value of x

(ii) each of the external angles.

SECTION B (60 Marks)

11. (a) The diagram below shows Mr. Kato¢s ground floor plan for his house. He wanted to buy a carpet to exactly cover this area.

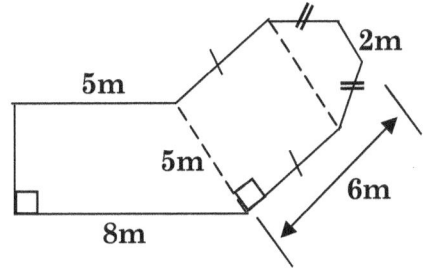

(i) Find the area of the carpet he needs to buy.

(ii) If a square metre of a carpet costs shs. 4500, how much money did he pay for this carpet.

(b) (i) Express 0.314 as a fraction.

(ii) Find the reciprocal of your answer expressed as a decimal.

12. In a club of 64 members, 34 like volley ball (V) and 28 like basket ball (B). the number of those who like neither of the two games is twice the number of those who like both of the games.

(a) Represent this information on a venn diagram

(b) How many members like both games

(c) How many members like neither of the games.

13. (a) Find the length of lines AB and CD, where: A (1, -3), B (-²/₃, 3), C (0, 0.5) and D (-4, 2 ½)

(b) Solve: $^{39}/_5 - \frac{1}{4} = a - {}^2/_5$

(c) Kayizzi obtained 42 marks out of 80 marks. Express his marks as a percentage.

14. Ali has twice as many books as Abbey. Abdu has 3 books less than the number of books Ali has.

   (a)   Write an expression representing the total number of books they have altogether and simplify it.

   (b)   If the total number of books they have altogether is 43. find how many books each has got.

15. (a) On a grid plot the points A(-3, 3), B (2, 1) C(1, 4) and D (-1, 4).

   (b) Join points A to B, B to C, C to D and D to A.

   (c) Name the figure ABCD obtained.

   (d) Through the line $y = 0$ as a mirror, draw the image of the figure ABCD and label it A'B'C'D'.

Extra Question

16. The table below shows the sex of students in S.1 at Blue stars academy.

Stream	Boys	Girls	Students
A	24	41	65
B	11	60	71
C	47	23	70
D	32	30	62
E	43	32	75
Totals	157	186	343

(a) How many streams are in S.1 class?

(b) How many boys are in S.1 class?

(c) What is the total number of students in S.1 class?

(d) Which streams have lesser girls than boys?

(e) Which stream has an almost equal number boys and girls?

MODEL TEST PAPER 9

SECTION A (40 Marks)

1. Work out:　　　(i) $450 - 4.561 + 23$　　　(ii) $\frac{3}{4} - \frac{2}{5}$

2. Given set　D = {All letters in the word development}

E = {All letters in the word employment}

(a) List the members for each set

(b) Find: (i) D ∩ E　　(ii) n(D ∩ E)

3. Complete the following sequences:

(i) 0.4, 0.6, 0.8, ……., ……..., ……..

(ii) $\frac{1}{2}$, $\frac{1}{4}$, $\frac{1}{8}$, ……..., ……..., ………

4. On the venn diagrams below shade;

(i) (A ∩ B)′　　　　　　(ii) (P ∩ Q′)

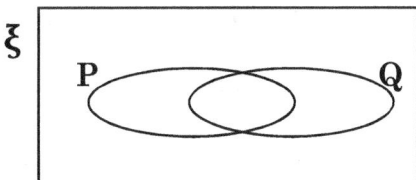

5. After a calculation, Ali got an answer 0.00034km². He was then asked to give his answer in m². What answer did he give?

6. Find the area of the figure PQRST.

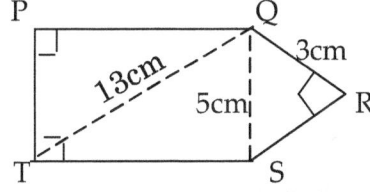

7. Simplify: $\dfrac{y(m-y)}{2p} \times \dfrac{8p^2}{y-m}$

8. Use a number line to evaluate: (i) $2 + (^-4) + 5 + (^-3)$ (ii) $5 + (4) + 8 + (^-6)$

9. (a) Find values of angles marked by letters in the figure below;

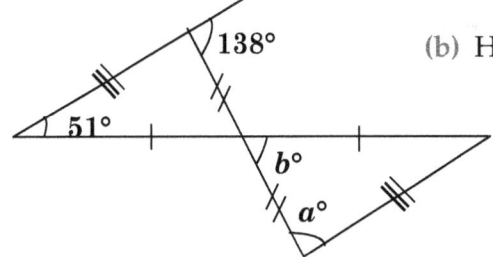

(b) How does the two triangles relate?

10. Write a word phrase for the expression: $\dfrac{n^2}{5} - 2 = {}^3/n$

SECTION B (60 Marks)

11. Given the venn diagram below, shade the following regions (draw a separate venn diagram for each illustration).

    (i) $A \cap B \cap C$ (ii) $(A \cap B \cap C)^{'}$ (iii) $(A \cup B \cup C)^{'}$

    (iv) $(A \cap B) \cap C^{'}$ (v) $(A \cap B)^{'}$

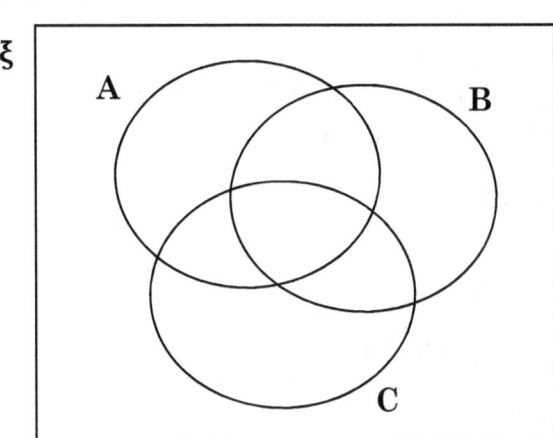

12.

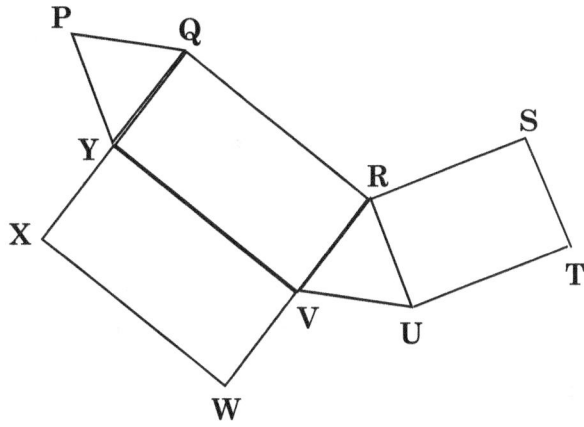

Above is a net that can be folded to form a solid.

    (a) Name the solid

    (b) Which letters will   (i) point U meet,   (ii) Point T meet

                                    (iii) Point R meet?

    (c) Which edge will meet edge: (i) QR   (ii) WX?

    (d) How many edges has the solid got ?

    (e) How many vertices has the solid got ?

13. The pattern below is made with matches.

 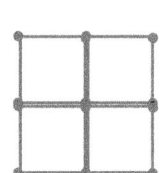 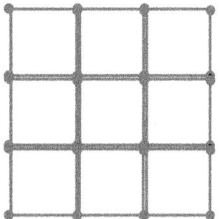

    (a) On a squared paper, draw the next diagram in the pattern.

    (b) Copy and complete the table below to represent the pattern.

Number of small squares	1	4	9	16	25	36
Number of matches used	4		24			

(c) Explain how to find the number of matches used for any diagram in the pattern. □

(d) How many matches are used when there are 225 small squares in the pattern?

14. Out of 50 students, 5 liked neither matooke nor cassava, 29 liked cassava but not matooke while a total of 7 students did not like cassava.

(a) Represent this information on a venn-diagram.

(b) How many did not like matooke

(c) How many did not like cassava

(d) How many liked both cassava and matooke?

15 (a) The perimeter of a square is 24cm.   Find: (i) the length of its sides
                                                    (ii) its area.

(b) On a grid: $-5 \leq x \leq 4$, $-6 \leq y \leq 5$

(i) Draw the lines $y = 2x$ and $y = 3x + 2$.

(ii) Give the coordinates of the point where these two lines meet.

Extra Question

16.

MODEL TEST PAPER 10

S E C T I O N   A   (4 0   M a r k s)

1. Fill in the missing numbers:

$$\begin{array}{r} 4\ \ 1\ \ 8 \\ +\ 1\ \square\ 6\ \square \\ \hline 8\ \ 2\ \square\ 7 \end{array}$$

2. (a) Write down two consecutive numbers that are not multiples of 3. Add them and note the answer.

   (b) Repeat (a) for several values (about 5 pairs of values)

   (c) What do you realise? Make a conclusion.

3. Solve for x. $8(2x – 3) + 21 = (7 + 6x)$

4. A rectangle has length 6x cm and width 5cm. If its area is 300 cm². Find:

   (i) the value of x (ii) the length of the rectangle and

   (iii) the perimeter of the rectangle.

5. 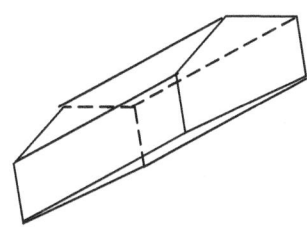 (a) What is the name of the figure shown?

   (b) How many edges has the solid got?

   (c) How many vertices has the solid got?

   (d) How many rectangular faces has the figure?

6. A car driver saw the following words in his mirror from a vehicle behind his car.   (a) What are these words?

   HƆAOƆ ƎƆI⅃OꟼꟼОⅬIⅭE ꓛΟАꓛΗ (b) What effect have these words under gone ?

7. If $q = -2$, $b = 0$, and $c = 6$, evaluate: (i) $3x – ac$ (ii) $\dfrac{3(102b – 7a)}{c}$

8. Express the number 38001.498 to: (i) the nearest whole number

   (ii) the nearest tenths

9. (a) Write down any two consecutive triangular numbers and find their sum.

 (b) Repeat (a) for several values

 (c) What do you realise? Make a conclusion.

10. Express: $\dfrac{32 \times 2^6 \times 128}{2^2 \times 8}$, as a single power of 2.

SECTION B (60 Marks)

11.

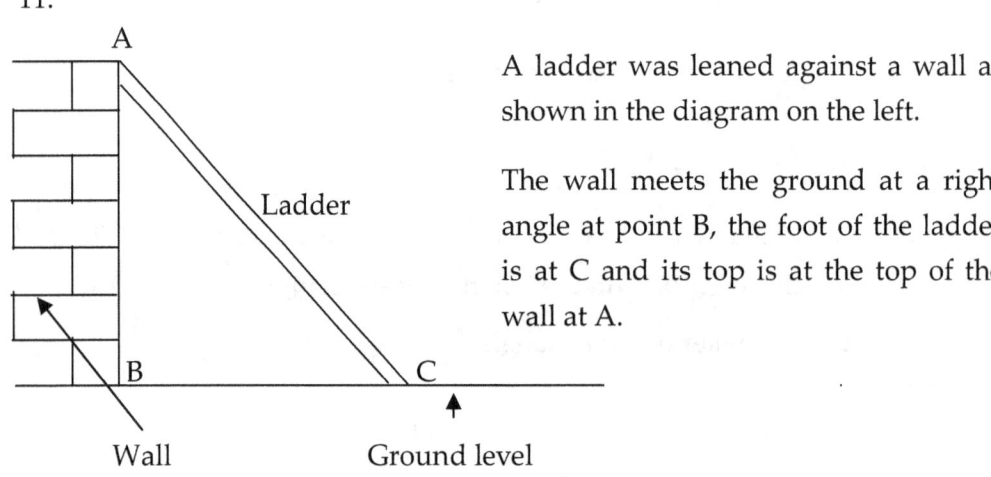

A ladder was leaned against a wall as shown in the diagram on the left.

The wall meets the ground at a right angle at point B, the foot of the ladder is at C and its top is at the top of the wall at A.

 (a) If the height of the wall (height AB) is 3.8m and the distance between the wall and the foot of the ladder (length BC) is 1.5m. Find the length of the ladder.

 (b) If the top of the ladder was adjusted to be 1.4cm below point A;

 (i) sketch the diagram representing this situation leaving points, A, B and C in their positions, but use A' and C ' for the new respective positions of A and C.

 (ii) Find the new distance between the foot of the ladder and the wall.

12. (a) Four books weigh $(x - y)$g, $(34 - 5x)$g, $(2y + 7x)$g and $(x - 3y)$g.

 (i) Write an expression for the total weight of the four books.

(ii) If the total weight of these 4 books is 834g and $x = 3y$, find the weight of each book in grams.

(b) Given the line $3x - 5y = 4$, tabulate the coordinates of points lying on this line for values of x such that $-5 \le x \le 5$.

13. Given the coordinates A(0, 4), B (1, 4), C (2, 5), D(3, 4), E(4, 4), F(3, 3) and G(1, 3).

(a) On a grid for values of x such that $-2 \le x \le 8$ and values of y such that $-4 \le y \le 5$, plot the points A, B, C, D, E, F and G.

(b) Join the points A to B, B to C, C to D, D to E, E to F, F to G, and G to A, to obtain figure. ABCDEFG.

(c) On the same grid, draw the line $y = x$.

(d) Through the line $y = x$ as a mirror line reflect the object ABCDEFG to obtain the image $A^{I}B^{I}C^{I}D^{I}E^{I}F^{I}$

14. Given the sets; N = {1, 2, 3, 4, 5, 6, ... 50}

$\qquad$ A = {1, 3, 6, 10, 15, 21}

$\qquad$ B = {all multiples of 3 less than 50}

$\qquad$ C = {all odd square numbers less than 50}

$\qquad$ D = {2, 3, 5, 7, 11, 13}

$\qquad$ E = { all square numbers}

(a) List the members of sets B and C

(b) What is the type for set E

(c) Find: $\qquad$ (i) n(B) $\qquad$ (ii) n(C)

(d) State *true* or *false* in the following:

$\qquad$ (i) $\quad$ 49 $\in$ C $\qquad\qquad$ (ii) $\quad$ B $\subset$ N

$\qquad$ (iii) $\quad$ D $\in$ C $\qquad\qquad$ (iv) $\quad$ 12321 $\in$ E.

15. (a) Kawalya obtained 34 out of 50 marks and Kato obtained 64 out of 80 marks.

  (i) Who of the two boys performed better than the other?

  (ii) What was their average percentage mark?

  (b) (i) Factorise 32.7 x 8 + 32.7 x 2, and hence evaluate to get the answer.

  (ii) Given the figure below, find its perimeter and area.

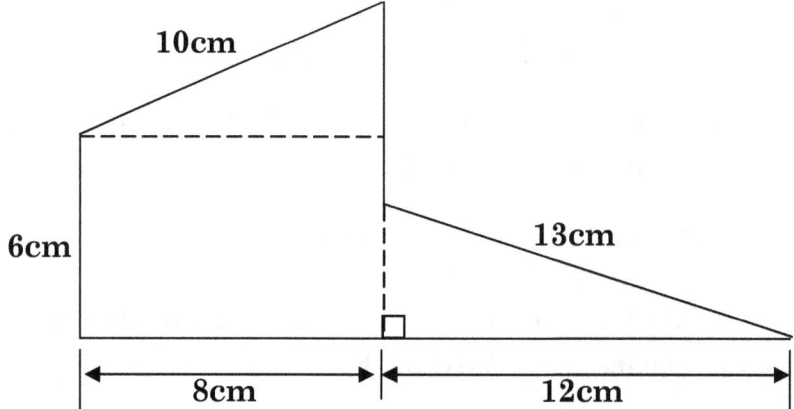

www.ingramcontent.com/pod-product-compliance
Lightning Source LLC
Chambersburg PA
CBHW081106170526
45165CB00008B/2342